The Brain's Light Switch

How Optogenetics Learned to Flash Individual Neurons On and Off — and What This Reveals About the Circuits of Memory, Fear, Addiction, and Mental Illness

Dr. Rosalind K. Thayer

Published by NobleTrex Press

For permissions and other inquiries, write to:

P.O. Box 3132, Framingham, MA 01701, USA

Contents

Introduction

What if a memory could be summoned with a flash of blue light?

Not metaphorically. Not in a movie. A real memory—or something close enough to make an animal behave as if the past has returned—triggered by a pulse lasting a few thousandths of a second.

That sentence sounds like science fiction written by someone who has never had to file a grant application. Yet over the past two decades, neuroscientists have learned to do something astonishingly close to it. They can insert light-sensitive proteins into selected neurons, guide light into the brain, and then turn those cells on or off with a precision that matches the brain's own electrical language. A pulse of light can make a mouse freeze in fear, seek a reward, stop a movement, retrieve a memory, or fail to retrieve one. The important point is not the theatricality. The important point is that for the first time, researchers can test whether a particular group of cells is not merely active during an experience, but actually helping to cause it.

This is the sort of advance that changes not only experiments, but arguments.

For more than a century, neuroscience has been haunted by a frustrating asymmetry. The brain is clearly physical tissue, made of cells, molecules, and electricity. But the phenomena we care about—memory, craving, panic, habit, grief, attention, relief—arrive as experiences. We feel them from the inside, while science must inspect them from the outside. Bridging those two perspectives has always required a special kind of audacity. You have to believe that fear is not a foggy essence but a pattern in a circuit. That a decision can be decomposed into signals arriving at cells at specific moments. That a psychiatric symptom, however intimate and human, might one day be explained with the same clarity that we bring to reflexes and vision.

That belief has often been justified. It has also often been oversold.

If you have ever read that scientists have found the brain region for love, the gene for depression, the circuit for addiction, you have encountered the brain sciences in one of their least careful moods. The problem is not that the brain has no organization. It very much does. The problem is that living systems are layered, tangled, and annoyingly good at doing several things at once. A neuron can be active while an animal is afraid without causing fear. A burst of activity may reflect a movement, a memory, an expectation, or a correction. A brain area famous for one function may contain dozens of interleaved cell types doing very different jobs. To say that some cells "light up" during a behavior is often just to say they were there when it happened.

Correlation, in other words, is cheap. Causation is expensive.

Optogenetics made causation less expensive.

The name sounds technical, and it is, but the basic idea is deliciously simple: borrow from nature a protein that responds to light, install it in neurons, and use flashes of light to control the flow of charged particles across the cell membrane. Since electrical charge is what neurons use to compute, communicate, and decide whether to fire, controlling that flow means controlling neural activity itself. Blue light can excite a chosen population. Yellow or green light can silence another. Red-shifted tools can reach deeper or more gently. Pattern the light carefully enough, and you can begin to ask not just whether a circuit matters, but when, how much, and in which cells.

The elegance of this is easy to miss because the method is now so widely discussed that it risks sounding inevitable. It was anything but inevitable. Optogenetics is one of those rare scientific developments that feels, in retrospect, as if several separate stories had been waiting decades to collide: the study of microbes and algae that navigate by light; the molecular physics of pigments that change shape when struck by photons; the engineering of viruses that can deliver genes into selected cells; the fabrication of tiny optical hardware sturdy enough to work in behaving animals; and a cultural shift in neuroscience toward asking harder causal questions. When Karl Deisseroth's group and, independently, Ed Boyden and colleagues showed in 2005 that neurons could be made to respond to microbial opsins with millisecond precision, it was not merely a new tool. It was a new standard of proof.

That matters far beyond the laboratory bench.

Take memory. Every one of us depends on the brain's ability to store traces of the past and reactivate them at the right moment. Yet memory often fails in ways that are painfully familiar. Names vanish. Traumas intrude. Addiction makes certain cues feel unnaturally important. Depression can turn recollection into selective evidence for hopelessness. At the center of all these conditions is a basic question: how does the brain assign meaning to experience and then recover it later? Optogenetics did not answer that question all at once, but it gave researchers a way to stop speaking in vague nouns and start testing mechanisms. Susumu Tonegawa's lab electrified the field when it showed in 2012 and 2013 that neurons active during the formation of a memory could be tagged and later reactivated to drive memory-related behavior. Suddenly the old, slippery concept of an engram—a physical trace of memory—looked less like philosophical speculation and more like something you could manipulate.

Or take fear. Fear is useful until it isn't. It keeps animals alive, yet in anxiety disorders and post-traumatic stress it becomes sticky, generalized, badly calibrated. For years, scientists spoke of the amygdala as though it were a singular fear center, a phrase tidy enough to fit on a magazine cover and misleading enough to cripple thought. Optogenetics exposed the shortcut. Different cell populations, different inputs, different outputs, different timings: all can push behavior in different directions. The amygdala did not turn out to be less important than we thought. It turned out to be more interesting.

The same is true of reward and addiction. Dopamine is one of the most famous molecules in biology, celebrated and blamed in almost equal measure. It has been called the chemical of pleasure, motivation, wanting, reinforcement, compulsion, and prediction error, sometimes all before lunch. Optogenetics helped sort out this conceptual traffic jam by allowing scientists to deliver brief, precisely timed activation to dopamine neurons and ask what follows. A burst at one moment can reinforce an action. Silencing a pathway at another can blunt cue-triggered seeking. The result is not a single grand theory but something better: a map of distinctions that behavior alone had blurred.

And then there is mental illness, where the stakes become both scientific and personal. Diagnoses such as depression, obsessive-compulsive disorder, schizophrenia, and addiction are still defined primarily by symptoms. That is necessary in the clinic, but symptoms are not mechanisms. Two patients may share a diagnosis while differing dramatically in the underlying circuit dysfunction that produced it. Optogenetics does not cure this problem, and it is not itself a treatment for most brain disorders. But it has transformed the way researchers think about them. It suggests that symptoms may emerge from particular patterns of communication among cell types and pathways, and that understanding those patterns is the missing bridge between molecular detail and lived experience.

If that sounds abstract, consider the practical implications. The person unable to stop checking the stove, the veteran who flinches at a sound that should be harmless, the adolescent whose reward system learns too eagerly from drugs and too weakly from ordinary life, the patient whose depression does not respond to medicines that alter brain chemistry in broad strokes—these are circuit problems in the plainest sense. The brain is not a sack of neurotransmitters. It is an organ of organized traffic. If you want to understand why one thought loops, one memory intrudes, one urge overpowers judgment, you eventually have to ask which signals are driving which cells, and when.

Optogenetics is the first technology that let neuroscience ask those questions with enough precision to be dangerous—in the best sense of the word. Dangerous to sloppy explanations. Dangerous to cartoonish diagrams of brain regions with single labels. Dangerous to the habit of mistaking what is easy to measure for what actually matters.

Of course, a "light switch" in the brain is never as simple as a switch on the wall. Real light scatters in tissue. Neurons differ in shape, excitability, and connectivity. The proteins used in optogenetics have quirks: some are fast, some slow, some conduct ions efficiently, some barely. Viruses spread unevenly. Genetic targeting is wonderfully powerful and constantly imperfect. Stimulating a cluster of cells may send activity ricocheting through networks in ways no one intended. The most exciting optogenetic experiments are therefore not magic tricks but carefully designed acts of restraint. They depend on controls, calibration, anatomy, recordings, and skepticism. A convincing result is rarely "we shone light and something happened." It is "we targeted this population, verified expression, measured the physiological effect, ruled out heating and other artifacts, compared timing conditions, and found that the behavior changed exactly as the circuit model predicted."

That combination of audacity and discipline is one reason the field is so fascinating. Another is that optogenetics reveals a deeper truth about science itself: revolutions often arrive disguised as hybrids. This story is not owned by neuroscience alone. It belongs equally to microbiology, biophysics, genetics, engineering, psychiatry, and the study of animal behavior. It began, in an important sense, with microbes and algae that had no interest whatsoever in helping mammals remember anything. Curiosity-driven work on how tiny organisms sense and move toward light uncovered the molecular parts list. Other researchers figured out how those proteins worked. Still others learned how to express them in neurons and deliver light into living brains. A field took shape because methods, ideas, and people from different corners of science became unexpectedly compatible.

That is one of the happiest patterns in modern research: a question asked for no practical reason at all can become the keystone of an entire new technology. The algae did not know they were preparing a toolkit for psychiatry.

As these pages unfold, we will move from the smallest scales to the most human ones. We will start with a problem that seems almost philosophical—how to tell whether a brain signal is causing a behavior or merely accompanying it—and show why milliseconds matter. We will build the minimum electrical intuition needed to understand what neurons are doing when they fire, and why a light-gated protein can hijack that process so effectively. We will meet the microbial opsins themselves, those improbable molecular machines that convert photons into ion flow, and see how the peculiar habits of single-celled organisms ended up reshaping the study of thought.

From there, the technical story becomes inseparable from the conceptual one. How do you get the right gene into the right cells? How do you send light deep into soft, scattering tissue without fooling yourself about what you actually stimulated? How do you measure what happened afterward? What counts as strong evidence when behavior is noisy and brains are complicated? These are not side issues. They are the difference between a clever demonstration and a trustworthy claim.

Only then are we ready for the scientific payoffs that made optogenetics famous: reactivating memory-linked ensembles, dissecting fear circuits, testing what dopamine bursts actually teach, parsing motivation from movement, and asking how patterns of neural activity might contribute to mood disorders. By the end, the technology reaches back toward medicine. The brain is a difficult place to shine light safely and usefully, but the retina offers a tantalizing exception, and efforts to restore vision have already pushed optogenetics toward the clinic. Alongside the excitement comes a necessary sobriety about hype, invasiveness, ethics, and the distance between making a mouse do something and helping a person.

If there is a single idea holding all of this together, it is that understanding the brain requires learning to write in it, not just read from it. For decades, neuroscience excelled at eavesdropping. Optogenetics supplied a controlled interruption. And once you can interrupt a conversation with precision, the hidden rules of the conversation begin to show themselves.

A flash of light, then, is not merely a trick for controlling neurons. It is a way of asking the brain a cleaner question. Are these cells enough? Are they necessary? Does timing matter? Does this pathway carry fear, value, memory, movement, or something more specific than any of those broad words suggest? Each pulse is a probe sent into the machinery of meaning.

That is why the subject matters even if you never plan to pipette a virus or align an optical fiber. The same circuits being teased apart in laboratories are the ones that allow you to learn from embarrassment, ignore a temptation, fall into a habit, wake from a nightmare, recognize a face, or feel your pulse jump when a sound resembles an old danger. The boundary between "basic neuroscience" and ordinary life is much thinner than it appears. What changes in the lab is not the material under study, but the sharpness of the questions.

And the sharpness is new.

For the first time in the history of brain science, researchers can pick out one defined population among millions of neighboring cells and ask, with near-native timing, what happens if these cells fire now. Not this region in general. Not this neurotransmitter in bulk. These cells. This projection. This pattern. This moment.

That is an extraordinary power. It deserves awe, caution, and a very close look.

Chapter 1

A Flash in the Dark: Why Controlling One Neuron Changes Everything

Brains do not reveal their secrets merely by lighting up on a scan or firing during an experiment; activity can accompany a thought, a fear, or a movement without causing any of them. What follows is the story of how a beam of light, aimed with exquisite precision, gave neuroscientists a way to ask a sharper question: if this exact set of cells is pushed or silenced at just the right instant, what changes—and what does that say about the hidden wiring of the mind?

1.1 When a Busy Neuron Tells You Almost Nothing

Imagine watching a city at night from an airplane. A neighborhood brightens; somewhere else, traffic thickens; a stadium flares like a small sun. You could make a decent guess about what's happening on the ground—a concert, a game, an accident—but you would also be wrong in ways that matter. Lights can rise because of celebration or because of a power surge. Traffic can snarl because of a parade or because a bridge is closed. Activity is information, but it is not yet explanation.

Neuroscience has lived with the same temptation for more than a century. Put an electrode near a neuron, or aim a scanner at a brain region, and something reliably happens: when an animal turns left, certain cells fire; when a person sees a face, certain patches of cortex become more active; when fear grips the body, certain deep structures show a burst of signal. The hard part begins right there, at the moment when the evidence looks most satisfying. A neuron can be *busy* during a behavior and still not be the reason the behavior happens.

Part of the problem is almost too ordinary. Brains are densely connected. If one node in a network becomes active, many others will follow, some because they are driving the process and others because they are being driven by it, echoing it, monitoring it, stabilizing it, or simply sharing the same incoming information. You can think of a neuron as sitting at a crowded intersection. When the city is awake, cars pass through constantly. Counting those cars tells you the intersection is active. It does not tell you whether the intersection is the destination, the source, or just a convenient route.

This is why some of the most impressive-looking findings in neuroscience can be so treacherous. A modern experiment might record from hundreds or thousands of neurons simultaneously and discover a crisp relationship between their firing and a choice an animal is about to make. Sometimes a computer can decode that choice from the neural activity before the animal moves. That can feel like mind reading. Yet decoding is not the same as proving that the brain *uses* that signal to make the choice. The recorded neurons might be downstream spectators receiving a copy of the decision, or they might be responding to subtle movements the experimenter did not notice (a change in posture, a micro-saccade, a shift in breathing). A signal can be predictive without being pivotal.

Even more seductive is the language of brain imaging, where it has become normal to say that a region "lights up" for love, pain, morality, or addiction. The phrase itself does some of the misleading work. The most common functional MRI signal (the BOLD signal, short for blood-oxygen-level-dependent) is not a direct measure of neurons firing. It reflects changes in blood flow and oxygenation that are linked, in complicated and state-dependent ways, to neural activity. In some circumstances, the BOLD signal can track the collective bustle of synapses and local processing more closely than it tracks the spiking output that neurons use to send messages to distant targets. A bright patch on a scan is closer to a heat map of metabolic demand than to a readout of commands being issued. So when a scan "activates," it may be telling you that a region is busy, not that it is in charge.

A famous prank-like demonstration made this painfully memorable. In 2009, a group led by Craig Bennett showed that if you apply sloppy statistics to fMRI data, you can find "significant" brain activation in a dead salmon. The point was not that fMRI is nonsense. The point was that the human desire to see patterns, combined with complex data and flexible analysis choices, can manufacture confident stories out of noise. The salmon was a warning sign: observation alone, even when it looks quantitative and high-tech, can flatter us into overclaiming.

Long before fMRI, the same tension existed in a more tactile form. Wilder Penfield, operating on epilepsy patients in the mid-twentieth century, electrically stimulated the surface of the brain and sometimes elicited vivid experiences: a smell, a sound, a flash of a remembered scene. These reports were astonishing. They also raised a question that is still with us: did Penfield stimulate the seat of the memory, or did he nudge a network in a way that caused the experience to be constructed elsewhere? Electrical stimulation spreads; it can recruit fibers passing through the area; it can perturb nearby inhibitory cells that normally keep activity in check. A result can be real and still be hard to interpret. You can evoke a percept without revealing the natural code that usually produces it.

So how do you move from "this neuron is active" to "this neuron matters"? The simplest answer is also the most demanding: intervention. Not just watching, but *doing*. If the neuron is genuinely part of the machinery, then changing its activity should change the outcome in a lawful way.

Two ideas structure this logic, and they are easy to say but difficult to earn: *necessity* and *sufficiency*.

A neural population is *necessary* for a behavior, in a given situation, if silencing it makes the behavior disappear or degrade in a specific way. If an animal can no longer express a learned fear when those cells are quiet, that suggests those cells are not mere bystanders.

A neural population is *sufficient* for a behavior, in a given situation, if activating it makes the behavior appear even when it otherwise would not. If turning on a particular ensemble causes a freezing response in a safe environment, that suggests those cells can drive the circuit toward fear.

Neither claim is absolute in the way everyday language makes it sound. Necessity can be masked by redundancy: the brain may have backups. Sufficiency can be artificial: driving cells with an experimenter's rhythm may produce an effect that resembles the natural behavior without proving that the brain normally uses that pattern. Still, necessity and sufficiency are the beginning of causal language in neuroscience, because they force the argument to survive an active test rather than a clever interpretation.

In practice, the necessity test often sounds more straightforward than it is. Suppose you silence a set of neurons and an animal stops performing a task. The first, alarming possibility is that you have not removed the task-specific computation at all; you have simply made the animal sleepy, anxious, distracted, or unable to move. Many behaviors are fragile. A mouse that feels slightly unwell will explore less. A person with a mild headache will respond slower. A circuit perturbation that changes arousal can masquerade as a memory deficit, and a circuit perturbation that disrupts motor coordination can masquerade as a decision deficit.

This is why careful experiments try to ask not just whether performance drops, but *how* it drops. Does the animal fail only when memory is required, or also on a simpler version of the task? Does it still see and move normally? Is reaction time affected, or only accuracy? Does silencing shift behavior in a specific direction (more risk-averse, more impulsive), or does it simply flatten all behavior into disengagement? The more specific and interpretable the impairment, the more credible the necessity claim becomes.

Sufficiency tests bring their own complications. A classic pitfall is that stimulating a circuit can produce a behavior by brute force, the way pulling a puppet's string produces a wave. That tells you something is connected, but it may not tell you what the circuit naturally represents. If you stimulate a motor pathway, you can cause movement. That does not mean those neurons encode the *desire* to move. It means they can drive muscles, directly or indirectly. Similarly, if stimulating a fear-related circuit causes freezing, you have shown that the circuit can recruit defensive behavior. You have not automatically shown that the circuit is a dedicated "fear center"; it could be part of a broader system for salience, surprise, threat, or negative value, depending on the context and the downstream state of the animal.

A deeper complication is that stimulation is not a neutral probe. Neurons are living cells with machinery that adapts. If you repeatedly activate them, you can change synapses and reshape the network. The intervention can become part of the learning, not merely a readout of it. This matters a great deal when interpreting experiments on memory and addiction, where the whole phenomenon involves plasticity—the brain's ability to rewire itself.

That brings us to the unglamorous backbone of causal neuroscience: controls. In an optogenetic experiment, the most obvious control is "light without the light-sensitive protein." If shining light into the brain changes behavior in animals that do not express an opsin, then you have discovered an artifact: heat, tissue damage, a stress response, or a visual cue the animal can detect. Light can warm tissue. Light can scatter into the eye. A tethered cable can alter posture and movement. Even a faint click from a shutter can become a cue. Without controls, a dramatic behavioral effect can be an expensive version of tapping the animal on the shoulder.

Controls also include "opsin without effective stimulation" (a sham pattern, a different wavelength, a blocked fiber), and, crucially, comparisons that match the animal's general state. If stimulation makes an animal freeze, is it because you have recalled a fear memory, or because you have made the animal feel unwell? If silencing abolishes a learned response, is it because you erased the memory, or because you interfered with the ability to express it? Those distinctions are not philosophical; they change what you would conclude about mental illness or addiction. A treatment that blocks expression is different from a treatment that repairs a memory trace, and different again from one that blunts arousal.

It helps to keep in mind a common logical trap: the brain can contain signals that are *real* and *reliable* and still not be the controlling variable. Consider a simple everyday example. A smoke alarm reliably correlates with smoke. You could predict a fire from its noise. But the smoke alarm does not cause the fire. It is a sensor. Some neurons are sensors in this sense: they report what is happening elsewhere. Other neurons are closer to actuators: they drive changes. Many are both, depending on the moment. Recordings often cannot tell you which role you are seeing, because the same firing pattern could be a readout, a command, or a byproduct of shared input.

Historically, one of the most influential examples of "busy but ambiguous" activity comes from the study of dopamine neurons. For decades, recordings in animals showed dopamine cells changing their firing with rewards and reward-predicting cues. That was a landmark discovery, central to modern theories of learning. But even there, correlation did not settle everything. Were dopamine spikes the teaching signal that updates predictions, or merely a reflection of something computed elsewhere? Intervention experiments—including those made far more precise by optogenetics—pushed the field from observation to causal tests by asking whether patterned dopamine activation can substitute for reward in driving learning, and whether silencing dopamine signals prevents certain forms of reinforcement learning. The more refined the intervention became in time and cell type, the closer researchers came to answering the question that recordings alone could not close.

A similar logic applies to fear, memory, and craving. When an animal freezes in response to a tone, neurons in the amygdala often show vigorous activity. When an animal navigates a familiar environment, hippocampal neurons can fire in location-specific patterns. When a person sees a cue associated with drugs, a network of regions becomes active. None of that, on its own, proves that those particular cells are the levers of the experience. Brains are full of correlated fireworks. The challenge is to identify which sparks are wiring and which are smoke.

Intervention, when done well, changes the basic grammar of the evidence. It turns a passive observation into an active query: if I remove this component, does the system fail in a specific way; if I add this pattern, does the system produce a specific outcome? Yet the phrase "remove this component" hides another difficulty. Neurons are not isolated resistors you can unplug. Silencing one population can release another from inhibition, causing it to become *more* active. Activating a group can trigger compensatory inhibition that suppresses neighbors. A small push can ripple through a network and return as an echo from somewhere else. Causation in circuits is rarely a straight arrow; it is often a loop.

This is one reason researchers hesitate to make sweeping statements like "this neuron causes fear." A more honest claim sounds narrower and, paradoxically, more informative: "under these task conditions, in this animal, at this moment, activity in this defined population is required for normal expression of the learned response," or "driving this population with this temporal pattern is sufficient to bias behavior toward avoidance." The price of precision is modesty in wording. The reward is that the claim can survive contact with competing explanations.

One final twist is worth holding close, because it explains why controlling even a small set of neurons can feel like a conceptual earthquake. In many engineered systems, a component's function is stable and context-independent. Flip a switch and the lamp turns on. In brains, context is part of the mechanism. A neuron's effect depends on who else is active, what the animal expects, whether it is hungry, whether it is stressed, and what has been learned. The same intervention can do different things on different days. That does not mean the tool failed. It means the brain is not a simple machine with labeled parts; it is a living network whose meaning is distributed and state-dependent.

A "busy" neuron is therefore an invitation, not a conclusion. It tells you where to look, not what to believe. The real test is whether changing that neuron's activity changes the animal's life in a way that rules out simpler stories—a test that demands not only clever technology, but also humility about how easily the brain can fool a confident observer.

1.2 Thought at the Speed of Lightning

The brain's basic currency is a brief electrical event called an *action potential*, often shortened to a *spike*. The name sounds dramatic, but the drama is mostly about speed. A spike is not a slow swell; it is a sharp pulse, commonly about 1 to 3 milliseconds long. If you blink, you miss it by a factor of a hundred. If you speak a syllable, you miss it by a factor of a thousand. A spike is closer in duration to the snap of a camera shutter than to any movement you can feel in your own body.

That matters because much of what brains do depends not only on *which* neurons spike, but *when* they spike. In everyday life, timing is the difference between catching a falling glass and watching it shatter. In the brain, timing is not merely a measurement; it is often part of the code.

A simple thought experiment makes the point. Imagine you are about to cross a street and a cyclist appears unexpectedly. Your eyes register motion; your brain updates its prediction; your muscles change course. The entire chain can unfold in a fraction of a second. Within that fraction, there are smaller deadlines. Signals from the retina reach deeper visual areas quickly; motor commands must arrive at spinal circuits in time to alter your step; the timing between excitation and inhibition in local networks can determine whether activity spreads or is contained. If you want to understand which neural events actually *cause* the decision to stop or go, you have to work at something like the brain's own pace.

This is the second old frustration of neuroscience, paired with the first. It is difficult enough to move beyond correlation and ask what is necessary or sufficient. But even when you can intervene, the intervention has to be fast enough to separate cause from consequence. A tool that acts slowly can tell you that a component matters in some broad sense while blurring the moment-to-moment causal chain into a fog.

For much of neuroscience's history, the interventions were blunt in time. Lesions are the extreme example: remove or damage a brain area and see what the animal can no longer do. Lesions can be informative, sometimes profoundly so. The classic neurological case of Henry Molaison, known for decades as patient H.M., showed that damage to structures in the medial temporal lobe can make it impossible to form new episodic memories while leaving many other abilities intact. That single life changed the science of memory. Yet a lesion is not a timed perturbation; it is a permanent amputation. It cannot tell you when, during a memory task, a structure contributes, or whether its role changes across learning versus recall, or whether its influence is direct or routed through other regions. It answers a different question: what happens when the brain is forced to live without a piece of itself?

Drugs are gentler but still slow. Give a chemical that blocks a receptor, and you can reduce activity in a circuit for minutes or hours. This can reveal that particular molecules matter for learning, anxiety, or attention. But drug effects rise and fall gradually, and they often spread. The brain is bathed in the bloodstream; receptors are reused across multiple circuits; a drug that quiets one kind of signaling in one region likely alters it elsewhere too. Even with careful targeting, pharmacology tends to change a whole *state*—arousal, anxiety, motivation—rather than a precise event in a precise moment.

Electrical stimulation, the workhorse of twentieth-century brain science and medicine, occupies an ambiguous middle ground. It is fast; you can deliver a pulse in milliseconds. But electricity does not respect cellular identity. A stimulating electrode can excite passing axons as well as local cell bodies. It can recruit nearby inhibitory neurons and excitatory neurons together. It can propagate backward along fibers and forward into distant targets, sometimes in ways that surprise even experts. The stimulation is precise in time but often indiscriminate in *what* it activates. That can be perfect if you are trying to treat a disorder by nudging a circuit—deep brain stimulation for Parkinson's disease has transformed many lives—but it is treacherous if you are trying to understand natural computation in fine detail.

The problem becomes clearer if you picture a symphony. Suppose you want to know whether the oboe line is essential for the emotional arc of a passage. A lesion is like removing the entire woodwind section and then trying to infer what the oboe used to contribute. A drug is like slightly detuning many instruments at once, then listening for broad changes. Electrical stimulation is like banging a piano key with a mallet: fast, loud, and not very selective. You may still learn something real. But you will struggle to make claims about the role of a single musician playing a specific phrase at a specific moment.

Neuroscientists have long known that timing must matter. The earliest careful measurements of spike timing, from work on peripheral nerves to the famous squid giant axon studies that established the electrical basis of the action potential, revealed just how rapidly excitable membranes can change. Later, recordings from cortex and deeper brain areas showed that neurons can alter their firing within tens of milliseconds after a sensory event, and that animals can make meaningful perceptual decisions astonishingly quickly. Even without knowing the details of a circuit, you can feel the constraint: if the brain makes a choice in, say, a couple hundred milliseconds, then the decisive neural events must occur well within that window. Any method that takes seconds to ramp up is like arriving at the scene after the verdict has already been reached.

This is why optogenetics landed with such force. It did not merely add another tool to the box; it changed what kinds of causal questions could be asked with a straight face. In 2005, Edward Boyden and Feng Zhang and their colleagues showed that a microbial opsin called channelrhodopsin-2 (ChR2), when expressed in neurons, could allow light to evoke spikes on a millisecond timescale. The phrase "millisecond-timescale" is easy to skim past, but it is the hinge of the whole story. It means the intervention can be paced like the brain's own language. It means you can turn a set of neurons on and off quickly enough to test whether spikes at a particular moment are doing causal work, rather than merely reflecting a slow change in mood or metabolic state.

To appreciate what that buys you, consider a fear response. A mouse hears a tone that has been paired with a mild shock and freezes. Many neural events unfold between tone and freezing: sensory processing of sound, retrieval of the learned association, evaluation of threat, recruitment of defensive motor programs. If you silence an area for ten minutes and freezing disappears, you have learned something useful: the area matters in some way. But you have not learned whether it matters for hearing the tone, recalling the association, generating fear, or executing freezing. Because the silencing blankets the whole period, it cannot tell you where the causal thread runs.

Now imagine you can silence with millisecond precision. You can switch the perturbation on only during the first 50 milliseconds after the tone begins, or only during the next 50, or only at the moment the animal transitions into freezing. If freezing disappears only when you perturb a narrow time window, the circuit's role becomes more legible. The experiment begins to resemble a careful edit rather than a demolition. You are no longer asking whether a structure is important in general; you are asking which neural events, at which times, are load-bearing.

The same logic applies to perception. In vision, for example, neurons in early cortex respond quickly to edges and contrast, while later areas integrate information into objects and meaning. If you intervene too slowly, you may knock out all of perception at once and conclude that everything you touched is "visual." With fast, timed perturbations, you can ask whether early spikes are essential for detection, whether later spikes contribute to recognition, and whether feedback signals from higher areas are needed for conscious report. Timing lets you separate stages that are otherwise piled on top of each other.

Timing also matters because brains are not steady-state devices. Many circuits operate through oscillations and rhythmic patterns: waves of activity that coordinate when different populations are most excitable. Even if a neuron fires the same *number* of spikes over a second, the *placement* of those spikes within tens of milliseconds can change which downstream neurons they influence. Synapses are not passive wires; they have dynamics and delays. Some inputs arrive when a neuron is poised to fire and have a large effect; the same inputs arriving slightly later might be ignored. A slow perturbation can easily wash over these temporal rules, while a fast perturbation can test them.

It is tempting, at this point, to treat optogenetics as a perfect stopwatch: light goes on, spikes happen; light goes off, spikes stop. Real neurons are never quite so obedient, and this is where the engineering details become part of the scientific logic. ChR2 is a remarkable protein, but it has kinetics of its own—a characteristic speed with which it opens, conducts ions, and closes. Those kinetics interact with the electrical properties of the neuron membrane. Under some stimulation regimes, ChR2 can produce extra spikes or prolonged depolarizations, especially if the light is strong or the pulses are long. In other words, the tool can shape the pattern you think you are imposing.

This is not a minor technical footnote; it is a reminder that "millisecond control" is not a label you can simply attach and forget. The difference between 5 milliseconds and 20 milliseconds is enormous at the level of spikes and synapses. Recognizing this, researchers engineered faster variants to tighten temporal fidelity. Ilana Gunaydin and colleagues (2010) demonstrated "ultrafast" optogenetic control using engineered opsins such as ChETA, designed to follow high-frequency patterns more reliably and reduce unwanted extra spiking. In circuit experiments where the question is whether neurons must fire in a particular rhythm or sequence, such improvements are not luxuries; they determine whether an experiment tests the intended hypothesis or a distorted version of it.

The need for speed is also a clue about what kinds of claims are plausible. Suppose someone says, "This region decides." If the evidence is a slow manipulation that shifts behavior over minutes, the claim is hard to interpret: perhaps the manipulation altered alertness, stress hormones, or general motivation. A decision, in the ordinary sense, is a temporally precise event: a commitment that flips one action into another. To argue about decisions, you want an intervention capable of matching the decision's timescale. Otherwise, the most you can safely say is that the region influences the animal's general tendency or bias, not the moment of commitment itself.

A similar caution applies in psychiatry and addiction, where the language of "circuits of depression" or "craving pathways" can suggest tidy switches. A depressed mood can last for weeks, but the neural events that compose it still occur in the same millisecond spikes as everything else. Fast tools allow you to test whether particular bursts of activity drive a feeling of aversion, whether certain patterns reinforce a habit, or whether brief interventions can interrupt a spiral of rumination. They also force a more honest kind of specificity: not just *where* and *what*, but *when*.

There is, finally, a human reason this matters. People experiencing panic, compulsions, or intrusive memories often describe them as sudden: an unwanted flash, a surge, a snap into a familiar groove. Those experiences can feel instantaneous and unstoppable. If the underlying circuits turn on in tens of milliseconds, then a tool that takes minutes to act will always arrive late. Speed does not guarantee a cure, but it changes the kinds of explanations that can be tested and the kinds of interventions one can imagine. The fantasy of a brain "light switch" begins, at least, with the recognition that the brain's own switches are fast.

The true lesson of millisecond neuroscience is not that the brain is a lightning bolt. It is that the brain is built from countless tiny lightning bolts, coordinated with exquisite timing, and that understanding it means matching that scale closely enough to see which flashes are doing the work.

1.3 Not Where in the Brain, but Which Cells

Point to almost any spot in the brain and you will find a crowded neighborhood. Even a region with a crisp name and a recognizable shape is less like a single building and more like a city block: different cell types, different wiring patterns, different "jobs" happening side by side. That simple fact is easy to forget when the evidence arrives as a glowing blob on a scan, or as a single electrode trace from a particularly chatty neuron. It is also the reason a great deal of brain localization talk feels both compelling and unsatisfying. Saying "the amygdala is for fear" or "the hippocampus is for memory" works at the level of dinner-party conversation, but it collapses under the weight of what those regions actually contain.

The shift from *where* to *which cells* did not happen because neuroscientists suddenly became pickier. It happened because, as soon as you can intervene with more precision, the brain starts answering different questions. A blunt perturbation encourages blunt conclusions. A precise perturbation makes blunt conclusions look wrong.

Start with a familiar example: imagine stimulating an entire brain region electrically. The effect can be dramatic, and it might even look interpretable—a movement, a pause, a defensive reaction, a change in mood. Yet inside that region live many populations with opposing influences. Some neurons are *excitatory*: when they fire, they tend to increase the chances that their targets will fire. Others are *inhibitory*: when they fire, they tend to decrease the chances that their targets will fire, often by opening channels that make it harder for the target cell to reach its own spike threshold. These are not philosophical categories; they are physical. The synapse—the microscopic junction where one neuron influences another—can be wired to push activity forward or to apply a brake.

If you jolt a region wholesale, you often recruit both pushers and brakers at once. You can activate neurons that send information outward and neurons that keep that information in check. You can drive local cells and axons merely passing through on their way to somewhere else. In a dense microcircuit, activating everything is like pressing the gas and the brake simultaneously while also grabbing the steering wheel: the car certainly does something, but it becomes difficult to infer what the driver normally intends.

This is why region-level explanations can mislead even when they are not exactly wrong. The hippocampus is deeply involved in memory, but it is not a memory monolith. Different subregions receive different inputs and send different outputs. Some hippocampal circuits are especially tuned to context—where you are, what the environment feels like, the layout of cues—while others carry signals that look more like time, or novelty, or prediction. The amygdala participates in fear, but also in learning what matters, and in linking sensory cues to good outcomes as well as bad ones. A "fear region" that is also active during reward is not a paradox; it is a hint that the region contains multiple interleaved components, and that the key variable is which subset you are measuring or manipulating.

Once you accept that regions are mixed, a more useful unit of explanation comes into view: the *microcircuit*. A microcircuit is a local network of cells connected in stereotyped ways: excitatory neurons sending signals to other excitatory neurons, inhibitory interneurons shaping the timing and gain of that excitation, and projection neurons carrying the output to targets elsewhere in the brain. Behavior then becomes less like a spot on a map and more like a pattern of influence: which cells talk to which others, with what sign (exciting or inhibiting), and under what conditions.

A historical detour helps here, because neuroscience has seen this movie before in another organ. Early maps of the heart, for example, could tell you where the muscle is thick and where the blood flows. But the heart's rhythm is not explained by geography alone. It depends on specialized cells in the sinoatrial node, pathways that conduct electrical waves, and timing relationships that coordinate contraction. The heart is a circuit problem inside an anatomical structure. The brain is the same, except the circuit has learned to represent the world and to change with experience.

One of the most striking demonstrations of "which cells" thinking came from memory research, and it arrived with an almost science-fiction premise: what if you could label the specific neurons that were active when a memory formed, and then later turn those same neurons back on?

The idea behind this approach, often called *engram tagging*, rests on a plausible intuition. When you learn something—a place where danger occurred, a cue that predicts reward—some neurons become active during that learning episode. Those neurons are not necessarily the memory itself, but many theories propose that a memory has a physical footprint in the pattern of strengthened connections among an ensemble of cells. An *engram* is the term often used for that physical substrate: the set of changes in the brain that make the memory persist and become retrievable.

Liu and colleagues (2012) made this logic concrete in a contextual fear paradigm. Mice learn to fear a context—a particular cage with particular smells and textures—when it is paired with a mild shock. The behavioral readout is freezing. The experimenters used a tagging strategy to label neurons in the dentate gyrus of the hippocampus that were active during the learning. Later, in a different context, they reactivated that tagged ensemble optically. The result was a recall-like behavior: the mice froze even without the original context cues. In the careful details lies the conceptual punch. Reactivating a different ensemble, tagged in a different context, did not produce the same freezing. The effect depended on *which cells* had been tagged by the learning episode.

This is a sufficiency test with teeth. It does not merely say "the hippocampus lights up during memory." It says that a specific subset of hippocampal neurons, defined by their activity during learning, can later drive a behavior that looks like memory expression. The memory is no longer anchored only to a region; it is anchored to an ensemble embedded in a circuit.

Yet even here the old interpretive anxieties return in a new form. Optical reactivation is powerful, but it can also be unnatural. It may synchronize neurons more tightly than natural recall does. It may drive them at a rhythm the brain would not normally choose. And because neurons that fire together can strengthen their connections, reactivation can itself change the engram. The tool that reads can also write.

Ramirez and colleagues (2013) made this unsettlingly clear with a twist that captured public imagination: *false memories*. Using an engram-tagging approach, they reactivated neurons representing one context while delivering shock in another, creating an association that did not match the animal's actual experience. The memory system can be tricked because, at the level of circuits, a memory is not a perfect recording of the past; it is a set of signals that guide future behavior. If you artificially supply those signals at the right moment, the brain can treat them as if they were evidence. The experiment is a marvel of causal control and a cautionary tale at the same time. If stimulation can implant an association, then a "reactivation" experiment might be probing recall, or it might be nudging the network into new learning. Often it is doing some of both.

That duality matters for any attempt to connect these findings to mental health. Intrusive memories in post-traumatic stress disorder, cravings in addiction, ruminative loops in depression: all can be understood as maladaptive patterns of retrieval and reinforcement. The hope is that identifying the responsible circuits might allow selective interventions. The warning is that interventions that seem to "erase" or "restore" could, in reality, be reshaping the system in ways that are hard to predict. The brain's memory machinery is not simply a library; it is also a rehearsal space.

The "which cells" view becomes even more illuminating when you realize that ensembles do not carry only factual content. They are embedded in networks that assign value. A memory is rarely neutral. A place can be safe or dangerous; a cue can be inviting or threatening; a social interaction can feel warm or humiliating. The emotional coloring of memory—its *valence*—is not an optional extra. It is often the point.

Redondo and colleagues (2014) showed that manipulating tagged ensembles could alter not only whether a memory is expressed, but the valence associated with it under certain pairing conditions. By reactivating ensembles while the animal experienced new reinforcement, they could bias whether the memory would later evoke avoidance or approach. The result hints at a deep principle: a memory trace participates in a broader circuit, and its meaning depends on where its outputs go and what state the downstream network is in. "The same memory" can lead to different actions if the circuit context changes.

At this stage, the story might seem to be about ensembles as if they were tiny islands of meaning: find the right group of neurons and you can toggle a behavior. But the brain does not store memories or emotions in isolated pools. Ensembles are nodes in a web, and what often matters is not only which cells fire, but where their axons go.

This is where circuit logic sharpens again. A single excitatory neuron might send branches to multiple targets, each with different consequences. Activating the cell body indiscriminately activates all its outputs at once, like turning on a building's main breaker and powering every appliance simultaneously. But sometimes a question is more specific: does a particular projection from region A to region B drive avoidance, while a different projection from the same region A to region C drives seeking? If you can control neurons based on their connections—projection-defined control, emphasized in Karl Deisseroth's circuit work and reviews—then the unit of explanation becomes a pathway rather than a region.

The emotional relevance of this is easy to feel. Consider addiction. A cue associated with a drug can trigger craving, but craving is not a single thing in the brain. It involves attention, motivation, memory, stress, and action selection. If you stimulate or silence a whole region and see craving-like behavior change, you may still be unsure whether you altered the remembered value of the cue, the impulse to act, or the ability to inhibit the impulse. Projection-specific manipulations promise a more surgical question: which route is carrying the drive, and which route is carrying the brake?

Even without describing the technical methods in detail, the conceptual move is plain. The brain is not a set of containers labeled with functions. It is a set of interacting populations linked by synapses. A given anatomical area is a tangle of microcircuits with different signs and destinations. If you want to explain memory, fear, or choice in a way that might eventually inform treatment, "where" is rarely enough. Two patients can have activity in the same region and yet be suffering from different circuit failures: one might have an overactive drive pathway, another a weakened inhibitory control pathway, another an overly strong memory association that keeps feeding the drive.

This is also why terms like "activation" can be misleading. When a region becomes more active, it may mean that excitatory neurons are driving a downstream target harder. It may mean inhibitory neurons are clamping down and suppressing output. It may mean both, in a push-pull rebalancing. The same net signal can hide opposite underlying mechanisms. A rise in blood flow to a region, or an increase in average firing rate, is like hearing that a crowd got louder. It does not tell you whether the crowd is cheering, booing, or arguing among itself.

Microcircuit thinking forces a more honest humility about single-neuron stories, too. A neuron that seems to "encode" a fear memory might simply be a relay conveying information from elsewhere. Another neuron might be silent most of the time but become decisive at a critical moment, shaping whether a defensive response is initiated or withheld. Some inhibitory neurons, though fewer in number, can have outsized influence because they control the timing and synchronization of larger excitatory populations. In a circuit, impact is not always proportional to how much a neuron fires.

The payoff of this perspective is not only scientific elegance; it is a way of speaking about mental life that respects its complexity without surrendering to mysticism. Fear, craving, and recall feel unitary from the inside, but the circuits that produce them are mixtures: excitation and inhibition, local loops and long-range projections, ensembles that carry content and pathways that carry value. The most consequential difference is often not whether a region is involved, but which cell types and which connections are active, and whether they are acting as accelerator, brake, or steering.

A brain region is a place. A circuit is a set of relationships. When relationships change, the same place can produce a different mind.

1.4 A Light Switch That Isn't Really a Switch

The phrase *turning neurons on and off* has the satisfying snap of a household light switch. It promises a clean world: a single lever, a single circuit, a predictable result. The appeal is obvious. If a memory, a fear, or a craving can be switched on by activating a particular set of cells, then perhaps the disorder is just a switch stuck in the wrong position. The danger is equally obvious: brains are not lamps, and neurons are not copper wire.

Optogenetics came as close as neuroscience has ever come to making the metaphor feel real. It did so by exploiting a biological trick that predates brains by billions of years: proteins that respond to light. In many microorganisms, light is not merely something to see; it is something to *use*. Some microbes swim toward it to harvest energy, others away from it to avoid damage. To do that, they evolved membrane proteins that absorb photons and respond by moving ions across the cell membrane. Those proteins are called *opsins*. In their native setting, an opsin helps a tiny organism navigate. In a neuron, it becomes a handle for controlling electricity with light.

A neuron is an electrical cell because its membrane separates charges. The membrane acts like a thin insulating barrier, and special pores called *ion channels* allow specific charged atoms—ions such as sodium, potassium, chloride, and protons—to cross. When ions flow, the voltage across the membrane changes. If the voltage reaches a threshold, the neuron generates an action potential, the millisecond spike that carries signals onward. So if you can control ion flow with light, you can control whether a neuron spikes.

Here is the basic sequence. First, genetic instructions for an opsin are delivered so that certain neurons manufacture that protein and insert it into their membranes. Second, light of an appropriate color is delivered to the relevant brain tissue, usually through a thin optical fiber or an implanted light source. Third, photons are absorbed by the opsin. Fourth, the opsin changes shape in a way that opens a pathway for ions. Fifth, that ion flow shifts the neuron's voltage, making spikes more or less likely depending on the opsin type. The entire chain can occur on the timescale that neurons themselves operate, which is why optogenetics became such a decisive tool for causal experiments.

Even in this simplified story, the metaphor of a switch is already straining. A household switch works because copper wire conducts electricity cleanly and predictably, and because the geometry of a circuit is fixed. In the brain, everything that matters is variable: the light spreads, the proteins vary in how strongly they are expressed, and the circuit being perturbed reacts.

Begin with the physics. Light entering brain tissue does not travel like a laser in air. Tissue is a complex mixture of water, lipids, proteins, and small structures at many scales. Photons scatter. They bounce, change direction, and gradually attenuate with distance. As a result, illumination tends to form a gradient: the tissue closest to the light source receives the strongest irradiance, and the intensity drops off with distance. This is not a technical nuisance; it changes what "precise" means. A fiber tip placed with sub-millimeter accuracy does not produce a sub-millimeter sphere of uniform stimulation. It produces a zone of strong light fading into weak light, and the boundary between "affected" and "unaffected" is not a crisp line.

Color matters, too. Different wavelengths scatter and are absorbed differently in tissue. Shorter wavelengths tend to scatter more, while longer wavelengths may penetrate farther but interact differently with opsins. The choice of opsin and the choice of light are linked, and the constraints are practical as well as biological: what can safely be delivered without heating tissue, what can be expressed robustly in the targeted cells, what can be driven at the desired speed.

Now add biology. Even if two neurons sit side by side, they may not have the same number of opsin molecules in their membranes. Gene expression is uneven by nature. Viral delivery, promoter strength, cell health, and developmental history all play a role. One neuron may be exquisitely sensitive, spiking with a whisper of light; its neighbor may hardly respond. If you increase light power to recruit the sluggish cells, you may overdrive the sensitive ones.

Overdriving has consequences. A neuron is not a passive recipient of current. If you depolarize it strongly for too long, it can enter *depolarization block*, a state in which the membrane becomes so persistently depolarized that the usual spike machinery fails to reset properly. From the outside, this can look like "silencing" even though you are shining light meant to activate. The same stimulation can excite at one intensity and suppress at another, purely through the neuron's own biophysics.

The opsins themselves are diverse, and that diversity matters for interpretation. Some opsins are light-gated *channels*, like ChR2, which allow cations to flow and tend to excite neurons. Others are light-driven *pumps*, often used for inhibition, that actively move specific ions. The switch metaphor becomes especially risky with inhibitory tools. Some inhibitory opsins are ion pumps that can create non-physiological ionic shifts. For example, moving chloride or protons across the membrane can change the internal ionic environment in ways neurons do not normally experience during everyday signaling. That can alter the chloride reversal potential, shift pH, or produce rebound effects when the light turns off. A behavioral change produced by inhibition is still a real causal effect, but it may not be equivalent to the natural process of inhibition in a microcircuit.

Timing, too, is less clean than the marketing slogans. Optogenetics can be fast, but temporal fidelity is not guaranteed by default. Opsin kinetics and membrane properties can yield extra spikes, plateau potentials, or altered synaptic release probabilities depending on the stimulation protocol. If the goal is to mimic a natural pattern of firing, the choice of opsin and the pattern of illumination become part of the hypothesis. The tool is not merely delivering a generic "on" signal; it is imposing a specific biophysical perturbation.

A concrete example makes this less abstract. In early optogenetic experiments, it was sometimes tempting to think that delivering 20 light pulses at 20 Hz simply forced a neuron to fire 20 spikes at 20 Hz. Often that is approximately true. But under some conditions, the same pattern can produce extra spikes or broadened depolarizations that change neurotransmitter release at synapses. Downstream neurons do not just count spikes; they are sensitive to bursts, pauses, and synchrony. A slightly distorted spike pattern can recruit a downstream circuit in a different way than the natural pattern would. Engineering faster or more temporally faithful variants, such as ChETA, was not merely about convenience; it was about keeping the causal test aligned with the biological question.

Then there is the largest complication of all: networks push back.

Turning on a neuron is never only about that neuron. It changes its targets. Those targets then change other cells, including inhibitory interneurons that may suppress activity locally. In many circuits, excitation quickly recruits inhibition. This stabilizes the network, preventing runaway activity. It also means that activating a set of neurons can trigger a wave of inhibition that suppresses other neurons you did not intend to affect. Conversely, silencing a population can release another from inhibition, producing a paradoxical increase in firing elsewhere. The behavioral outcome can reflect a circuit-level rebalancing rather than a direct line from stimulated cells to action.

This is why causal claims in optogenetics must be bounded. The honest conclusion often sounds narrower than the popular headlines: *in these conditions, perturbing this defined population changes this behavior or physiology*. The claim is still powerful. It is just not the same as saying, "this neuron is the switch for fear," or "this pathway is the addiction button." The brain rarely offers single buttons. It offers levers connected to many springs.

The need for controls becomes clearer in this light. An optogenetic experiment lives or dies by showing that light alone is not responsible for the effect. That means opsin-negative light controls: shining the same light into animals or cells that do not express the opsin, to test for heating, tissue damage, or sensory cues. It also means power titration: mapping how the behavioral or physiological effect changes with light intensity, watching for suspicious thresholds or saturations. Histology matters, too: confirming where the opsin is actually expressed and where the fiber actually sits. In vivo electrophysiology—recording neural activity during stimulation—provides a reality check on what the light is doing to spikes in the targeted population and beyond.

These practices may sound like the usual belt-and-suspenders of careful lab work, but they serve a deeper purpose. They prevent the switch metaphor from smuggling in an assumption of simplicity. They force the experimenter to treat light as a physical intervention delivered to biological tissue, not as an abstract command.

There is a historical echo here in the early days of electrical brain stimulation. When researchers first discovered that stimulating certain sites could evoke movements or sensations, it was tempting to identify those sites as "centers" for those functions. Over time, the field learned that stimulation is an intrusion that recruits fibers of passage, spreads through conductive tissue, and interacts with ongoing activity. Optogenetics improved specificity dramatically by restricting sensitivity to genetically defined cells, but it did not repeal the basic lesson. Every perturbation has a footprint, and the footprint must be understood.

The switch metaphor is still useful, as long as it stays tethered to reality. It is useful because it captures a genuine breakthrough: for the first time, it became possible to make rapid, genetically targeted perturbations to neural activity with light, aligning intervention timing with the millisecond scale of spikes. It is useful because it reminds us that the brain is causally mechanistic: push here, something changes there. It is useful because it makes the experimental logic of necessity and sufficiency feel tangible rather than philosophical.

Where it fails is in implying that the brain's causal structure is a set of isolated toggles. Light does not land on a single cell like a fingertip. It spreads and attenuates. Opsins do not behave like ideal relays. They have kinetics, ion selectivity, and downstream consequences. Circuits do not respond like simple appliances. They adapt, compensate, and reshape themselves.

A better metaphor might be less comforting: optogenetics is not so much a light switch as a way of introducing a precisely timed *perturbation* into a living network, then watching how the network reorganizes and what behavior emerges. Sometimes the perturbation is small and revealing, like tapping a crystal glass and hearing its resonant tone. Sometimes it is large and transformative, like striking a chord that the system then learns to repeat.

In that sense, optogenetics is both a microscope and a finger on the scale. It lets researchers ask causal questions that were previously out of reach, but it also demands a new kind of restraint. The brain will answer the question you truly asked, not the question you meant to ask, and the difference often lies in the details of light, protein, and circuitry. The most convincing experiments are the ones that treat those details not as obstacles, but as part of the explanation itself.

Chapter 2

Electric Cells, Chemical Whisperers: How Neurons Become Switchable

A neuron is not a tiny wire but a restless living battery, forever balancing leaks, charges, and chemical pressures at its surface. Once that hidden tension is clear, the logic of optogenetics snaps into focus: add a new gate for ions, and a flash of light can tip thought itself.

2.1 Voltage as a Living Resource

A neuron's most valuable possession is not a neurotransmitter or a gene. It is a difference. On one side of its membrane sits a soup of ions—charged atoms such as sodium (Na^+), potassium (K^+), and chloride (Cl^-). On the other side sits another soup, similar but not the same. The membrane is a thin oily barrier that does not let these charged particles drift freely. Because the two soups are held apart, and because the membrane is selectively porous, a tiny separation of charge builds up across that barrier. That separation is called the *membrane potential*: an electrical voltage between the inside and outside of the cell.

The phrase "resting potential" can mislead, as if the neuron is idle, as if nothing is happening until a thought arrives. The truth is closer to a bustling city at night seen from an airplane: the lights are steady, but the steadiness depends on constant work. Even when a neuron is quiet, ions are always leaking, pumps are always pushing, and energy is always being spent to keep the imbalance from collapsing. Rest is an achievement, not a default.

The inside of a typical neuron sits at around -70 millivolts relative to the outside. That number looks small, and it is small compared with the 120 volts in a wall socket. Yet for a membrane only a few nanometers thick, -70 millivolts is an intense electric field. The cell has, in effect, charged its own membrane like a biological capacitor: two conductive fluids separated by an insulator. Only a vanishing fraction of ions must shift position to create this voltage, but the *permission structure* for ions—which ones are allowed to move, and when—makes that voltage an exquisitely controllable resource.

A helpful way to picture the situation is to treat ions not merely as chemical ingredients but as crowds with different preferences. Potassium is more concentrated inside neurons than outside. Sodium is more concentrated outside than inside. Chloride tends to be more concentrated outside as well, although its exact balance depends on the neuron's type and developmental stage. These are not minor differences. They are steep gradients, more like reservoirs held behind dams than like gentle slopes.

Left alone with open gates, the gradients would run down. K^+ would tend to leave the cell (from high concentration inside to lower outside). Na^+ would tend to enter. Cl^- generally tends to enter, pulled by concentration and, depending on the membrane voltage, by electrical forces. If all these ions could move equally easily across the membrane, the inside and outside would quickly approach the same composition, and the voltage would fade toward zero. A neuron with no voltage is not dead instantly, but it is in trouble: it loses the poised electrical difference that makes it responsive, and it begins to fail at basic tasks of volume control and signaling.

Neurons prevent that leveling-out with two kinds of molecular machinery that play complementary roles: *leak channels* and *pumps*. The leak channels are protein pores that allow specific ions to pass more readily than others. At rest, the membrane is typically much more permeable to K^+ than to Na^+, largely because of K^+ leak channels. So potassium has an "easy exit," and it begins to drift outward. But as positively charged K^+ leaves, it makes the inside more negative. That growing negativity pulls K^+ back in electrically. Eventually, the chemical push outward and the electrical pull inward come into balance, and net potassium flow slows. This balance point is not chosen for convenience; it is a consequence of the existing gradients and the membrane's selectivity.

Sodium complicates the picture in an important way. Even at rest, some Na^+ leaks inward. The neuron's resting voltage is therefore not simply the potassium balance point; it is a compromise produced by multiple ions moving through multiple pathways. The key intuition is that resting potential reflects which doors are cracked open *right now*. Open more sodium-permeable doors and the cell becomes less negative. Open more potassium-permeable doors and it becomes more negative. Open chloride-permeable doors and the effect depends on where chloride's own equilibrium lies, but the common result is often stabilizing: chloride channels tend to pull the membrane voltage toward chloride's preferred level.

Leak channels, however, are only half the story. If leaks were all that existed, the gradients would slowly erode. The neuron would drift toward equality and electrical silence. Preventing that erosion is the job of pumps, especially the sodium—potassium pump, a molecular machine that consumes cellular fuel (ATP) to move ions *against* their gradients. In its typical cycle it exports three Na^+ ions and imports two K^+ ions. This does two things at once. It restores the gradients that leaks and electrical activity would otherwise dissipate, and it contributes a small direct electrical effect because it moves more positive charge out than in.

The result is a non-equilibrium steady state: the membrane potential is steady not because nothing happens, but because two opposing flows are matched. If that seems abstract, consider a leaky inflatable mattress with an electric air pump attached. The mattress can stay firm all night, but only if the pump runs. Firmness is not a natural resting state; it is a maintained state. A neuron's "firmness" is its voltage, and its pump is not optional.

This is why the brain is metabolically expensive. Much of the energy you burn while sitting quietly is used by neurons and their neighbors simply to keep ion gradients steep. The bill comes due because every meaningful electrical signal spends a little of that stored imbalance. A neuron that fires repeatedly, or receives heavy synaptic bombardment, is like a city during a heat wave: the infrastructure must work harder just to keep conditions stable.

A historical detail makes the living nature of this battery hard to forget. In the 1930s and 1940s, Kenneth Cole and colleagues developed ways to measure the electrical properties of nerve membranes with increasing precision, turning what had been an evocative metaphor—"animal electricity"—into a quantitative subject. Their work, and later the squid-axon experiments of Alan Hodgkin and Andrew Huxley, depended on a crucial practical fact: the membrane behaves like an electrical element whose properties can be measured and perturbed, but it is also a piece of living matter whose state changes with metabolism, temperature, and time. When early electrophysiologists struggled with recordings that drifted or nerves that "ran down," they were encountering, at the bench, the price of maintaining a biological voltage. The signal was never separable from the cell's ongoing labor to remain excitable.

Once the idea of a maintained voltage feels real, an important second idea becomes intuitive: *small currents matter*. The resting membrane potential sits in a range where a modest nudge can have consequences. Neurons are not perched at the very edge of catastrophe, but many are poised close enough to a critical boundary that tiny changes can decide whether they remain silent or generate an electrical impulse. That boundary is often called *threshold*. It is not a painted line on the membrane; it is the point at which the cell's own voltage-dependent channels begin to amplify a change rather than damp it. Before that amplification begins, the neuron behaves like a leaky integrator: inputs push the voltage, and the leak pulls it back.

Why would evolution build cells that operate near a tipping point? Because it is efficient. A neuron that required huge currents to respond would be deaf to the gentle, distributed signals of its neighbors. A neuron that were permanently over threshold would be a noisy fire alarm. Poising neurons near threshold allows networks to be sensitive without being unstable. It also allows computation by addition and subtraction: many small influences can be combined, so that no single input must carry the whole burden.

A concrete everyday example of "small currents matter" appears in medicine, sometimes dramatically. Changes in blood potassium can shift the excitability of nerve and muscle. Too little potassium outside cells (hypokalemia) can make the resting potential more negative and muscles weaker; too much potassium outside (hyperkalemia) can make cells less negative and, paradoxically, lead to dangerous electrical problems, including in the heart. The point is not the clinical details; it is that moving ions around in the body by what seems like a modest amount can shift whether excitable cells are ready to fire. Life depends on guarding the gradients.

Now consider the membrane itself as a physical object. Its lipid core resists the passage of charged particles, making the membrane a good insulator. That insulation is what allows voltage to exist at all. Yet the membrane is studded with protein channels—specialized gates that let certain ions cross. The neuron is therefore a controlled leak. That phrase can sound like an insult until one realizes it is the whole design: insulation plus carefully regulated pores creates a system in which tiny changes in permeability can redirect ion flow and therefore voltage.

This is the key engineering logic behind making neurons switchable. If a neuron's electrical state is built from the selective movement of ions through protein gates, then adding a new gate changes the rules. It does not matter whether that gate is native to the neuron or introduced by genetic engineering. If the gate opens, ions move; if ions move, voltage changes; if voltage changes, the neuron's probability of firing changes. The membrane does not care whether the instruction to open a pore comes from a neurotransmitter, a change in voltage, a drug, or a photon. It only "cares" about conductance: how easily ions can flow.

Optogenetics exploits this indifference. A light-sensitive channel inserted into a neuron is, conceptually, another leak pathway—but one under the experimenter's control. When light opens such a channel, the neuron's membrane potential can be driven toward a new value determined by which ions the channel passes. If the channel allows positive charge to enter (for example, allowing Na^+ or other cations through), it tends to depolarize the cell, making the inside less negative and often moving it closer to threshold. If the light-driven protein instead pumps ions or opens a pathway that pushes the voltage in the opposite direction, the cell becomes harder to excite.

Either way, the control knob is the same variable the neuron already uses: ionic permeability.

The "battery" metaphor becomes even more useful when one adds one more detail: batteries have internal resistance, and so do neurons. A neuron with many open leak channels has low resistance; current flows easily, and voltage changes are harder to produce. A neuron with fewer open leaks has high resistance; the same tiny current produces a larger voltage change. This is why the same optogenetic stimulus can have different effects in different cell types or even in the same cell under different states. The cell's current-to-voltage conversion depends on how leaky it is at that moment.

The living battery also has a kind of memory, though not the long-term memory of experiences. It has a short electrical memory set by the membrane's capacitance and resistance. A brief input does not vanish instantly; it changes the voltage and then the voltage relaxes back toward rest over a characteristic time. This matters because it means that inputs can summate: one small nudge followed quickly by another can add up to something decisive. If the second nudge arrives too late, the first has already faded. The neuron's physical properties therefore create a natural time window for integration, shaping what counts as "together" in neural time.

At this point it helps to hold three images in mind simultaneously, because neurons truly are all three. They are chemical bags with a boundary. They are electrical devices with stored charge. And they are energy-consuming machines that must continually pay to keep their resources stocked. The resting membrane potential is where those identities meet.

The crucial point is that voltage is not merely a readout of neural activity. It is a stockpile. The gradients of Na^+, K^+, and Cl^- are like money in different currencies, and the membrane's channels are the exchange booths. Open the right booth at the right moment, and current flows. That flow changes the voltage. And because the neuron's own proteins respond sharply to voltage, a slight shift can be amplified into an electrical event that travels and triggers chemical release downstream.

This is why "controlling neurons" is not a futuristic idea grafted onto biology from the outside. Neurons are already built to be controlled—by each other. Synapses change conductances; conductances redirect ion flow; ion flow moves voltage. Optogenetics simply supplies an additional conductance change, one whose trigger is light. In a cell that lives by guarding its electrical imbalance, any new gate is a new lever.

A last, human-scale way to appreciate how alive this resource is comes from what happens when the resource runs low. When oxygen is cut off during a stroke, neurons cannot make ATP efficiently. Pumps slow. Gradients begin to collapse. The resting membrane potential drifts. Neurons can depolarize en masse, releasing neurotransmitters in abnormal floods and swelling as ions and water move where they should not. The tragedy is not only that signaling fails; it is that the very mechanism that makes signaling possible—the maintained imbalance—is expensive enough that it cannot survive long without support. The brain's electrical readiness is purchased minute by minute.

Under normal conditions that purchase is so reliable that we forget it is happening. Yet every thought depends on it: billions of membranes, each one holding a negative interior against a salty world, each one leaky, each one continually repaired by molecular pumps, waiting for the tiniest current that changes the odds of firing.

2.2 The All-or-None Firecracker

A faint click on a camera shutter does not produce a faint photograph. Once the shutter mechanism catches, the whole sequence runs: blades move, light reaches the sensor, and the moment is captured. Neurons behave in a similarly decisive way. Below a certain point, you can push and prod their voltage and get only small, proportional changes. Cross the critical line, and the cell responds with a fast, stereotyped electrical discharge—a spike—whose size and basic shape are remarkably consistent from one event to the next.

That consistency is not a quirk; it is a design choice built into the proteins that sense voltage. A graded world is good for measuring, but a digital world is good for transmitting. A spike is closer to a yes/no pulse than to a slowly varying dial. The brain can still represent shades of meaning, but it often does so through the *timing* and *pattern* of spikes, rather than the amplitude of each one. The difference between a neuron firing now versus 10 milliseconds later can matter as much as the difference between whispering and shouting.

The word *threshold* makes it sound as if the neuron consults a rulebook and then decides to fire. In reality, threshold is the point where physics and protein kinetics create a positive feedback loop. A neuron's membrane contains channels that open and close depending on voltage. Some of the most important are voltage-gated sodium channels. When the membrane becomes slightly less negative, some of these channels open. When they open, sodium ions rush in, pulled by both the electrical attraction to the negative interior and the concentration gradient that keeps sodium higher outside than inside. That inward sodium flow makes the inside less negative still. And that further depolarization opens more sodium channels. The cycle feeds on itself.

A chain reaction has begun.

Positive feedback has a signature feel in everyday life: the microphone that squeals when it gets too close to the speaker, the rumor that accelerates once enough people repeat it, the snowball that finally gathers enough mass to roll downhill on its own. In neurons, the "speaker" is the sodium channel and the "microphone" is the membrane voltage. Below threshold, the system damps disturbances. Past threshold, it amplifies them. The switch is not perfect—there is no single voltage that is always the threshold in all conditions—but the transition is sharp enough that spikes behave like a distinct kind of event rather than merely a larger version of a small voltage wiggle.

Yet sodium channels are only half the story. A spike would be useless if it were a permanent state, like a firecracker that lights but never stops burning. Neurons need a way to end the event quickly and reset the membrane so that another spike can occur later. That role is played largely by voltage-gated potassium channels, along with sodium channel inactivation (a built-in shutoff). As the cell depolarizes and the spike rises, potassium channels begin to open with a delay. Potassium then flows outward, driven by its concentration gradient in the opposite direction from sodium. Outward potassium current pulls the membrane back down toward its resting negativity, repolarizing the cell. Meanwhile many sodium channels enter an inactive state from which they cannot immediately reopen, preventing the spike from instantly restarting itself.

The result is a pulse with a beginning, middle, and end on the scale of a few milliseconds. It is this speed that makes spikes ideal for communication through long axons. A slow voltage change would leak away and blur with distance; a regenerative spike is actively rebuilt along the axon, like a line of falling dominoes where each domino supplies the force to topple the next. The neuron spends its stored ionic gradients to create a signal that is difficult to ignore.

Much of what is known about this choreography comes from experiments that have the atmosphere of practical ingenuity rather than grand theory. In the mid-twentieth century, Alan Hodgkin and Andrew Huxley worked with the giant axon of the squid, an unusually large nerve fiber that could tolerate being impaled with electrodes and subjected to careful measurement. In 1952 they produced a quantitative model of the action potential that reconstructed the spike's waveform from the time- and voltage-dependent behavior of sodium and potassium conductances. The achievement was not just explanatory; it was predictive. It showed that the spike's shape could be generated from a few interacting currents—a leak, a fast inward sodium current with positive feedback, and a slower outward potassium current that restored order. A biological firecracker, but one whose fuse and blast could be written as equations.

There is a human subtlety to this work that tends to get lost in textbook retellings. To measure the currents that underlie the spike, Hodgkin and Huxley had to hold the membrane voltage at controlled values and watch what currents appeared. That required methods of "clamping" the voltage in place, fighting the neuron's own tendency to run away into a spike. In other words, understanding the spike demanded temporarily *preventing* it. The experiments were a tug-of-war with the very positive feedback that makes spikes possible.

Why does any of this matter for optogenetics? Because optogenetics rarely needs to create a spike from scratch. It needs to tip a neuron's already-poised membrane into the self-amplifying regime. The light-gated conductance is not a replacement for the cell's sodium channels; it is a nudge at the door that triggers the cell's own machinery to throw that door wide open.

That simple statement hides an important constraint: the nudge must be delivered with the right *speed* and *size*. If an excitatory opsin opens slowly or produces too little inward current, the neuron may depolarize a bit and then leak back toward rest without ever igniting the sodium-driven runaway. If it opens robustly, the membrane potential crosses threshold and the cell does the rest, producing a full spike whose timing is now locked to the moment of illumination.

This is why timing is not a decorative detail in experiments that "turn neurons on." A neuron does not respond to light like a lamp that brightens smoothly with a dimmer switch. Light can move the membrane voltage gradually, but spiking is a threshold event. That threshold makes the relation between illumination and output sharply nonlinear: small differences in membrane state, or in synaptic input arriving at the same moment, can decide whether the same light pulse produces silence or a spike.

A concrete example helps. Consider a neuron in the cortex that is receiving a constant background of synaptic chatter, keeping it hovering near threshold. A brief flash of light that opens cation channels might be enough to push it over, yielding a spike with millisecond precision. Now take the same neuron during a different brain state—perhaps drowsiness, anesthesia, or simply a moment when inhibition is stronger and the membrane is slightly more negative. The identical flash may yield only a small depolarization, no spike at all. The optogenetic "on" command is therefore not a simple digital instruction sent to a passive receiver. It is an extra current injected into an already-living electrical device whose responsiveness depends on what it has been doing in the preceding tens of milliseconds.

The all-or-none nature of spikes also clarifies why neuroscientists care so much about *where* spikes begin. In many neurons, spikes typically initiate near the axon initial segment, a region with a high density of voltage-gated sodium channels. The soma and dendrites may experience large graded voltage changes, but whether those changes are strong and fast enough to trigger the sodium channel cascade at the initial segment is what determines the final outcome. Optogenetic conductances placed in the soma or dendrites can influence this, but their effect will depend on how the voltage spreads within the cell and how quickly the current reaches the spike initiation zone. A light-gated channel that is perfectly capable of passing current can still fail to drive spikes reliably if its current is diluted by cellular geometry and leak before it reaches the crucial trigger point.

Once a spike occurs, the neuron briefly becomes a different kind of object. For a short period afterward, it is harder—sometimes impossible—to trigger another spike. This is the *refractory period*, a consequence of sodium channel inactivation and the lingering openness of potassium channels. Refractoriness is often described as a recovery time, but it has computational consequences. It limits how fast a single neuron can fire, and it turns spike timing into a resource that cannot be spent infinitely. It also helps spikes propagate in the correct direction along an axon: the region that has just fired is temporarily less excitable, encouraging the wave to move forward rather than backward.

For optogenetic control, refractoriness sets a ceiling on how faithfully a neuron can be made to follow rapid trains of light pulses. If light pulses arrive faster than the cell's ability to reset its sodium and potassium channel states, spikes will begin to fail or become irregular. A neuron that fires perfectly to 10 light pulses per second may start missing pulses at 40 per second, and it may respond with complex patterns at higher rates. These failures are not sloppiness in the method; they are the neuron's own protective rules made visible.

The same rules explain an apparent paradox: stronger stimulation does not always produce "more" spiking in a simple linear way. With electrical stimulation, too much current can recruit passing axons and adjacent cells, producing widespread effects. With optogenetics, too much depolarization sustained for too long can inactivate sodium channels, pushing the neuron into a depolarized state where it is actually less able to generate discrete spikes. The cell is not a loudspeaker that simply plays louder when the volume knob is turned. It has operating regimes, and the spiking regime sits between them: too little drive yields no spikes; too much steady drive can turn the spike generator sluggish.

A second paradox is even more instructive: you can sometimes silence a neuron by exciting it. This can happen when sustained depolarization inactivates sodium channels, producing a kind of block. It is an extreme illustration of the principle that spikes are not the same thing as depolarization. Spikes depend on a rapid, regenerative rise in sodium conductance. Depolarization is merely voltage. Optogenetics can control voltage directly by adding conductance, but whether that voltage becomes spikes depends on the cell's voltage-gated machinery being in the right state to respond.

These subtleties are precisely why the spike is such a useful event for the brain. All-or-none signals are robust to noise. If a spike is produced, it has enough amplitude to travel and enough stereotypy to be recognized. The price is that the brain must represent graded quantities in other ways. One common strategy is *rate coding*: stronger input makes a neuron fire more spikes per second. Another is *temporal coding*: information is carried in exactly when spikes occur relative to oscillations or to other spikes. The all-or-none spike supports both strategies. It is a token that can be counted, timed, synchronized, delayed, and compared.

Timing becomes especially vivid when one thinks about coincidence. Suppose a neuron receives two small excitatory inputs, each too weak to trigger a spike alone. If they arrive nearly together, their depolarizations can add, crossing threshold and triggering a spike. If they arrive separated by a longer interval, the first fades before the second arrives, and the neuron stays silent. A spike, in this setting, is the neuron's announcement that something happened within a narrow time window. That is a primitive form of computation, built not from logic gates but from leaky membranes and fast channels.

Optogenetics can exploit that property with unusual precision. A brief flash of light can be used as an artificial "input" whose timing is known to the millisecond. In a behaving animal, that allows a researcher to ask: if this set of neurons is nudged at this exact moment, does the circuit tip toward one decision rather than another? The all-or-none spike is the reason such questions can be sharp rather than vague. If light merely warmed the cell a bit or changed its chemistry slowly, the effect would be smeared in time and difficult to tie to a specific neural event. Because spikes are fast and thresholded, a fast optical conductance can produce a precise causal punctuation mark in the circuit's ongoing conversation.

There is another emotional aspect to the all-or-none spike that is easy to overlook because the language of neuroscience is so technical. Spikes are the brain's *commitments*. A neuron can receive thousands of synaptic inputs and remain undecided, its membrane potential wandering up and down in response. The moment it spikes, it stops weighing and starts broadcasting. In many neurons that broadcast leads, within a millisecond, to chemical release at synapses onto other cells. A spike is therefore not merely an internal event; it is a public act. It is the point at which private voltage becomes social influence.

That is why thc exact requirements for triggering spikes matter so much. An excitatory opsin that passes current into a neuron is effectively offering it a push toward committing. The push must arrive within the right time window and with enough strength to cross the positive-feedback boundary. If it does, the resulting spike is not a small local perturbation; it is a decisive signal that can propagate down an axon, recruit synaptic release, and ripple through a network. If it does not, the light may still shift voltage and modulate the neuron's readiness, but the outcome will be subtle, graded, and often dependent on whatever else is happening in the circuit at the same time.

The firecracker metaphor is useful not because spikes are destructive, but because they are sudden and standardized. Strike the match weakly and nothing happens; strike it well and the whole event follows. The brain relies on those sudden events to carve time into meaningful units. Milliseconds become the currency of coordination, because a spike is a millisecond-scale pulse that can align many cells, separate one sensory event from the next, and allow circuits to distinguish "together" from "apart." A brief flash of light, delivered to the right ion gate, can therefore do something surprisingly dramatic: it can decide *when* a neuron speaks.

2.3 Synapses: Excite, Inhibit, Compute

A spike is a crisp signal, but it is also a private one unless it reaches someone else. Neurons are social cells. Their influence comes from thousands of small contacts where one cell can change the odds that another will fire. Those contacts are synapses. The synapse is less like a wire than like a carefully engineered border crossing: a presynaptic terminal holds chemical packets of transmitter, an action potential arrives, and suddenly a microscopic negotiation begins between electricity and chemistry.

The basic choreography is quick. A spike invades the presynaptic terminal and changes its voltage. That voltage change opens channels that admit calcium ions (Ca^{2+}) from outside the cell. Calcium is a kind of molecular permission slip. Its influx triggers vesicles—tiny membrane-bound spheres packed with neurotransmitter—to fuse with the presynaptic membrane and release their contents into the narrow synaptic cleft. The transmitter molecules diffuse across that cleft and bind to receptors on the postsynaptic cell. Many of those receptors are ion channels themselves. When they open, they briefly change the postsynaptic cell's membrane conductance: they provide a new route for ions to flow, and therefore a new way for voltage to change.

This is the moment where the earlier discussion of voltage as a resource pays off. A synapse does not usually *inject* a fixed amount of voltage into a neuron. It changes conductance and thereby alters the voltage that results from existing ion gradients. Synaptic transmission is therefore deeply contextual. The same synapse can have a larger or smaller effect depending on what the membrane voltage was just before it opened, and on what other conductances are open at the same time. Chemical whisperers, yes—but whisperers who can sometimes trigger firecrackers.

The postsynaptic response to a single presynaptic spike is typically a small, graded change called a postsynaptic potential. If it depolarizes the cell, it is an excitatory postsynaptic potential (EPSP). If it makes the cell less likely to fire, it is an inhibitory postsynaptic potential (IPSP). These potentials are not all-or-none. They resemble ripples: a few tenths of a millivolt to a few millivolts, rising and decaying over milliseconds to tens of milliseconds. A single EPSP rarely triggers a spike by itself in the intact brain. What matters is the arithmetic of many EPSPs and IPSPs summing together across time and across the neuron's branching dendritic tree.

Excitation and inhibition are not moral categories; they are electrical effects. An excitatory synapse typically uses glutamate as its transmitter. Glutamate receptors that produce fast signaling are often ionotropic receptors, meaning that the receptor is itself an ion channel. When glutamate binds, the channel opens and allows positively charged ions to enter the neuron. That inward positive current pulls the membrane voltage toward a depolarized reversal potential, making a spike more likely.

Inhibitory synapses in many parts of the brain typically use GABA as their transmitter. Many fast GABA receptors are also ionotropic channels, often allowing chloride ions through. Opening a chloride channel tends to pull the membrane voltage toward chloride's reversal potential. Frequently that reversal potential is more negative than the spike threshold, so the neuron is driven away from firing. But inhibition does not need to be strongly hyperpolarizing to be powerful. It can work by changing *resistance*.

This is the idea of *shunting inhibition*, and it is one of the most counterintuitive tricks neurons use. Imagine trying to fill a bathtub while the drain is open. You can pour in more water, but the water level does not rise much because the drain provides a low-resistance escape route. In the same way, when inhibitory synapses open channels they increase the membrane's conductance and lower its input resistance. Excitatory currents arriving at the same time then produce a smaller voltage change, because the current is "shunted" through the open inhibitory conductance. The membrane potential might barely move, not because inhibition has pulled it far negative, but because inhibition has made voltage harder to change.

This conductance-based view of synapses is crucial for making sense of circuit manipulation. Turning on a group of inhibitory interneurons in a cortical circuit, for example, may do more than simply drive target neurons down in voltage. It can clamp their responsiveness, shorten the time window during which excitation can have an effect, and stabilize a network against runaway firing. Conversely, silencing inhibition can have effects that are larger and stranger than "more excitation," because it removes both hyperpolarizing forces and the shunting that normally controls gain.

Synapses also come in two main temporal styles: fast and slow. Ionotropic receptors produce fast postsynaptic conductance changes, well-matched to the millisecond timing of spikes. Metabotropic receptors, by contrast, act through intracellular signaling cascades. They can alter channels indirectly, change synaptic strength, and reshape excitability over longer timescales. A metabotropic influence is less like a whisper and more like adjusting the acoustics of the room: the same spoken word lands differently afterward.

A neuron's dendrites turn all these influences into computation. Dendrites are not passive cables. Their branching structure means that the same synaptic input placed in different locations can have different effects on the soma and the spike initiation zone. As the work of Greg Stuart and Nelson Spruston emphasized around 2000, synaptic integration depends on location, on timing, and on the electrical properties of the dendrite in between. A synapse on a far dendritic branch might produce a large local EPSP, yet only a smaller, filtered version reaches the soma. Meanwhile, other dendritic mechanisms can amplify or normalize signals, depending on the cell type and its state. The dendrite is where a neuron can behave less like a simple summing point and more like a tree of interacting subunits.

Spatial summation and temporal summation are the basic arithmetic operations. Spatial summation means that inputs arriving at different places on the dendritic tree add together as their voltage changes spread. Temporal summation means that inputs arriving close together in time add before the first has decayed away. Both forms of summation are shaped by conductance. If a neuron is in a high-conductance state—lots of synaptic channels open, lots of leak—each individual EPSP may be smaller and briefer, but timing can become sharper. If a neuron is in a quiet, high-resistance state, a single EPSP can produce a larger voltage change that lasts longer, but the membrane may be more susceptible to being pushed into spiking by accidental coincidences. The brain toggles between these regimes as it shifts between sleep, wakefulness, attention, and many other modes.

If synapses are arithmetic, inhibition often supplies the parentheses. One of the most common motifs in cortical circuits is *feedforward inhibition.* In this arrangement, an incoming excitatory pathway activates both a principal neuron (such as a pyramidal cell) and an inhibitory interneuron. The interneuron, being fast and often closer to threshold, fires quickly and delivers inhibition to the principal neuron with a slight delay. The result is a narrow time window: excitation arrives, and then inhibition closes the gate. The principal neuron can spike only if excitation is strong enough and well-timed enough to beat the inhibitory curtain. Jeffery Isaacson and Massimo Scanziani's 2011 review highlighted how such inhibitory patterns shape cortical responses, sharpening timing and preventing excitation from smearing into long, indiscriminate firing.

This timing window is not an abstract feature. It matters to perception and action. A sensory input that reaches cortex can evoke a precise volley of spikes whose timing carries information. Feedforward inhibition can make those spikes occur in a brief burst and then stop, preserving temporal contrast. It can also make the system more selective: only the best-matched inputs get through before inhibition clamps down. In a sense, inhibition converts raw excitation into a decision.

There is also *feedback inhibition*, where principal neurons excite interneurons that then inhibit the principal neurons in return. This creates a stabilizing loop, like a thermostat. When principal neurons become too active, inhibition rises to counterbalance them. It can prevent runaway excitation and can generate rhythms and oscillations that organize the timing of spikes across populations.

A third motif, *disinhibition*, looks like a double negative: an inhibitory neuron inhibits another inhibitory neuron, thereby releasing a principal neuron from suppression. Disinhibition can act like a gate that opens only under certain conditions, allowing a circuit to switch modes quickly. In some contexts, disinhibition is how attention is implemented: not by pouring in excitation everywhere, but by selectively lifting inhibition where processing should be enhanced.

All these motifs matter for a practical reason: a tiny intervention at one point can have large downstream consequences. Imagine a circuit in which a small group of interneurons supplies feedforward inhibition that normally constrains a principal neuron's firing to a narrow window. If those interneurons are silenced, the window widens. The principal neuron may now spike in response to inputs that previously would have been too late or too weak. That changes what the principal neuron communicates to its targets, which changes what *they* do, and so on. A manipulation that acts on a few cells can therefore redirect an entire cascade of spiking across a circuit.

This is why optogenetic tools that target inhibitory neurons can produce effects that feel disproportionate. A brief light pulse that activates interneurons can rapidly shunt excitation across a local network, suppressing spiking without needing to hyperpolarize every neuron dramatically. Conversely, a brief light pulse that silences interneurons can release a network from its normal constraints, leading to heightened firing and sometimes synchronized bursts. The circuit's behavior reflects its wiring and the balance of conductances, not merely the number of neurons being directly illuminated.

A concrete example brings the abstractions down to earth. Consider the difference between tapping the brakes and cutting the engine in a moving car. Activating inhibition is often like tapping the brakes: the system is still driven by ongoing excitation, but its output is restrained and timed. Shunting inhibition in particular is like adding friction at a specific moment, making other pushes less effective. Silencing inhibition can be like releasing the brakes entirely; the same excitatory inputs now have far more leverage. In both cases, the intervention can be local, but the consequences propagate because the circuit is interconnected and already active.

Synapses also introduce variability and plasticity, which matter for both computation and interpretation. The amount of transmitter released by a single presynaptic spike is not perfectly fixed. Vesicle release is probabilistic, and synapses can facilitate or depress depending on recent history. A synapse that has just been used heavily may release less transmitter on the next spike (depression), or it may release more (facilitation), depending on the synapse type and calcium dynamics. This means that the same presynaptic spike train can have different postsynaptic effects depending on context. The brain's computation is therefore not only about which neurons fire, but about the dynamic strength of their connections on the timescale of milliseconds to seconds.

When people first learn that neurons communicate chemically, it can sound slow or imprecise, like sending messages by drifting perfume. Yet synapses are fast enough to preserve millisecond timing because ionotropic receptors can open quickly and because the synaptic cleft is tiny. The chemical step does not erase temporal precision; it translates it into conductance changes. The impressive part is not that chemistry can be fast, but that it can be *tunable*. Synapses can vary in strength, location, receptor type, and timing. They can be modified by experience. They can be gated by neuromodulators that change mood, alertness, or motivation. The same basic mechanism—release, bind, open—can implement an enormous range of computational behaviors.

The push and pull of excitation and inhibition also helps explain how the brain can be both sensitive and stable. Sensitivity demands that small inputs can matter, that a faint sensory cue or a subtle memory trace can influence what happens next. Stability demands that the system not explode into uncontrolled firing. Inhibition is the stabilizer, but it is also a sculptor. It shapes when spikes occur, which inputs are allowed to count as coincident, and how strongly a neuron responds to the same excitatory drive. Excitation supplies the energy; inhibition supplies the form.

One last image helps hold this together without slipping into jargon. Think of a crowded conversation at a dinner party. Excitatory synapses are people leaning in, encouraging someone to speak. Inhibitory synapses are not simply people saying “be quiet.” Often they are people redirecting attention, narrowing the topic, or changing the timing so that only the most relevant comment lands. A spike is the moment someone finally speaks aloud. The conversation’s direction depends less on any one speaker than on who encourages whom, who interrupts, and who makes space at exactly the right moment.

In that setting, a small, well-timed intervention can change everything. A single person asking a pointed question can pull a discussion onto a new track; a single interruption can prevent a story from being told; a brief lull can allow a shy voice to enter. Neural circuits work the same way. Synapses are the micro-acts of permission and restraint that determine which spikes occur, and those spikes are the tokens that carry information onward. Computation emerges from countless such negotiations, each one grounded in a fleeting change in conductance across a membrane only nanometers thick.

2.4 Circuits Are Hypotheses You Can Break

A drawing of a circuit can be comforting. Arrows march from a sensory region to a decision region to a motor region, and it feels as if the mystery has been tamed. Yet a circuit diagram can also be a way of postponing the hard part. It can smuggle in an assumption: that the arrows represent causes rather than coincidences.

Brains are noisy and crowded. When a mouse freezes in fear, many regions change their firing; when a person craves a drug, many systems shift at once. If you only watch, the best you can usually say is that some neurons *tend* to be active when a behavior occurs. That is valuable, but it is not a circuit claim in the strong sense. A circuit claim is a wager about causation: this particular population, active in this particular pattern, is a reason the behavior happens.

A causal claim is fragile in a useful way. It can be broken. That is the difference between a story and a hypothesis.

The experimental verbs that break circuit hypotheses are surprisingly plain: *turn on* and *turn off*. The subtlety lies in what those verbs mean in biology, and in what counts as a fair test.

One way to phrase a circuit hypothesis is in the language of *sufficiency*. Suppose someone proposes that activity in a specific pathway—say, neurons that project from one brain region to another—is enough to generate a behavioral output. A sufficiency test asks: if those neurons are driven into activity, does the animal behave as if the natural trigger were present? If a flash of light makes a mouse freeze as though it heard a predator, or makes it seek a place as though it felt safe, that looks like sufficiency. The pathway may be acting like a command, or like a strong bias that can dominate other influences.

A different way to phrase the same hypothesis is in the language of *necessity*. Here the claim is that the pathway is required for the behavior under the conditions being studied. A necessity test asks: if that pathway is silenced at the relevant moment, does the behavior fail to appear, or does it weaken in a predictable way? If an animal no longer freezes to a fear cue when a certain cell type is briefly inhibited, that looks like necessity. The pathway might be an essential link in the chain, or a gate without which the rest of the circuit cannot do its job.

It is tempting to treat sufficiency and necessity as opposites, or as a checklist. In practice they are complementary angles on the same problem, and each can mislead if taken alone. A pathway can be sufficient without being necessary. If you stimulate the motor cortex strongly enough, a limb will move; that does not mean motor cortex is the only route by which movement can be generated. Conversely, a pathway can appear necessary in a narrow experimental setup and yet be bypassed in other circumstances. Brains are adaptable; they can recruit alternate routes, and they can change strategy when a familiar option is blocked.

Optogenetics made these verbs far sharper than they had been. Karl Deisseroth (2011) emphasized that introducing light-sensitive conductances allows millisecond-scale control that matches the timing of spiking. Gero Miesenböck (2011) stressed a complementary point: circuit diagrams are hypotheses, and optogenetics turns them into hypotheses that can be falsified by precisely targeted perturbations. The crucial move is that optical control is not merely fast; it can be selective. With the right genetic targeting, an experiment can aim at a defined cell type, or even at a particular projection, rather than at whatever happens to be near an electrode tip.

That selectivity matters because the brain is built from mixed neighborhoods. In a tiny volume of cortex, excitatory and inhibitory neurons interlace; axons from distant regions pass through; local cells send branches in every direction. Traditional stimulation methods often recruit a confusing combination of elements. Light can be delivered locally, but the key is that the *light-sensitive gate* is delivered only to the cells that embody the hypothesis. The perturbation is therefore not only spatially focused; it is biologically addressed.

A concrete example makes the logic vivid. Imagine trying to understand a fear circuit. A simple diagram might say: threat cue $\rightarrow$ amygdala $\rightarrow$ freezing. That cartoon can be refined into something testable: a particular population in the amygdala, when active, drives downstream targets that organize freezing. The sufficiency question is whether artificially driving that population can produce freezing without the threat cue. The necessity question is whether silencing that population prevents freezing in response to the cue.

These tests can be emotionally unsettling in their clarity, because they make behavior look programmable. In one common design, a mouse hears a tone that has been paired with a mild foot shock on earlier trials. The tone alone later evokes freezing, a stereotyped defensive stillness. If a researcher activates a defined amygdala population at the moment the tone would normally occur, and the mouse freezes in a similar way, the activation has substituted for a natural trigger. The sensation, for the reader, is sometimes a flicker of unease: if freezing can be evoked by a light pulse, how much of fear is a circuit reflex?

Yet the more sobering lesson is methodological. The light pulse does not prove that the activated neurons are *the* fear. It shows that under these conditions their activity can *drive* a fear-like output. That is already a strong claim, but it is not the whole story. Stimulation can be like grabbing the steering wheel: you can force a turn without proving that the car would have turned that way on its own.

Necessity tests address that concern, but they come with their own traps. If silencing prevents freezing, it is tempting to call the silenced population the indispensable fear node. But silencing can disrupt a circuit in many ways: it can remove a needed signal, it can unbalance excitation and inhibition, or it can change the timing relationships that allow downstream neurons to integrate inputs. A failure of behavior after silencing therefore supports necessity, but it does not automatically identify the missing computation. It identifies a broken component.

This is why the phrasing "circuits are hypotheses you can break" is more than a slogan. Breaking something can be informative, but only if you know what kind of break you made. A missing gear is different from a jammed gear, and biology has both. With optogenetics, one can sometimes choose perturbations that are briefer, more precise, and more interpretable than older methods, but interpretation still requires humility.

One reason optogenetics lends itself to causal language is that it works through the same biophysical currency neurons already use: conductance. A light-sensitive channel is not a foreign command; it is an added gate for ions. Turning it on is like adding an extra synapse that you control; turning it off (with a different tool) is like removing conductance that would otherwise be present. That continuity makes the intervention feel "physiological," yet it also means optogenetic effects can propagate through natural circuit motifs. Activating an excitatory population can recruit inhibition through feedforward and feedback loops; silencing an inhibitory population can disinhibit a network, indirectly increasing firing in many places. A perturbation targeted to one cell type can therefore echo widely, and sometimes the most dramatic change occurs in cells that never express the opsin at all.

Such echoes are not a defect; they are often the point. When a tiny population is thought to gate a behavioral state—a craving episode, a panic response, a compulsive ritual—the hypothesis is exactly that its influence is disproportionate. Circuit neuroscience is full of these proposed levers: small ensembles whose activity can tip a larger system. Optogenetics is one of the few methods capable of asking whether those levers are real.

The price is that optogenetic experiments need controls that are as carefully engineered as the perturbations themselves. Light is not an abstract variable; it is energy delivered to tissue. It can heat tissue. It can activate endogenous light-sensitive molecules weakly. It can scatter and reach unintended regions. The implants used to deliver light can damage tissue and provoke inflammation. Opsins themselves can have side effects: altering membrane properties, changing ionic loads, or producing unnatural firing patterns if driven too strongly or too long. Network-level consequences can also be distal or unintended: perturbing one node may recruit compensatory mechanisms elsewhere.

A particularly sharp issue arises when recording electrical signals at the same time as delivering light. Metal electrodes can produce photoelectric artifacts: brief voltage transients created not by neurons but by photons hitting the electrode or its surroundings. A naïve experiment can mistake a light-induced artifact for a spike, especially when the artifact is tightly time-locked to illumination—which is exactly the feature one hopes for in a successful manipulation. Artifact-aware design, careful choice of materials, and rigorous controls are therefore not bureaucratic hurdles; they are the difference between causation and illusion. One simple safeguard is to deliver identical light patterns to opsin-negative tissue and verify that any behavioral or electrophysiological effects vanish. Another is to check whether putative spikes have the expected waveform and refractory properties, rather than appearing as identical, instantaneous deflections.

Careful experimental logic also guards against a subtler interpretive error: confusing a pathway's ability to *evoke* a behavior with its normal role in *selecting* that behavior. An artificial stimulation pattern might be unlike anything that occurs naturally. If it drives freezing, that shows that downstream machinery is capable of producing freezing when sufficiently activated. It does not yet show that the same pathway carries the natural code for threat. More refined experiments try to reproduce plausible firing patterns, or to stimulate only at moments when the neurons would naturally be active, using recordings to guide the timing. The goal is not to make the animal do something odd, but to test whether plausible activity is enough.

The logic becomes especially powerful when activation and silencing are paired with a behavioral readout that is sensitive and specific. In addiction research, for example, one might measure lever pressing, place preference, relapse-like seeking after abstinence, or the vigor of pursuit under cost. In depression-related studies, one might assess social interaction, escape behavior, or reward sensitivity. In each case, the behavior is not merely a theatrical endpoint; it is the assay that turns circuit perturbations into claims about function.

A good circuit hypothesis therefore has an internal spine: it specifies *which* cells, *when* they should be active, *where* they project, and *what* downstream computation is predicted. Optogenetics does not magically supply those specifications. What it supplies is the ability to test them with unusually crisp interventions. A pathway is not "involved" in fear or reward in some vague sense; it either can, when driven, bias the system toward a measurable outcome, or it cannot. It either is required in a specific context, or the behavior persists without it.

Near the end of an experiment, there is a moment that feels like a small intellectual thrill: you press the button and the animal's behavior changes on cue. The thrill is not simply power. It is the feeling of a diagram turning into a lever, of an idea turning into a test, of a hypothesis being given a chance to fail.

When these experiments are done with care, the result is not a simplistic claim that "this neuron causes fear" or "that pathway causes addiction." The result is something more precise and more useful: under specific conditions, manipulating a defined component reliably changes the probability and timing of a behavioral outcome. That is the level at which circuit models become durable. They can survive being poked, contradicted, refined, and sometimes overturned—not because they are neat, but because they are breakable.

Chapter 3

Nature's Photoreceptors: The Proteins That Learned to Feel Light

Light seems effortless—mere brightness, color, and warmth—but living cells must somehow convert a passing photon into a physical act. Here the reader discovers how biology solved that problem with molecular devices so elegant that a trick first used by microbes to chase or flee the sun eventually became the foundation for switching brain cells on and off with flashes of light.

3.1 When Color Carries Force

Color usually arrives in conversation as decoration. A sunset is *red*, a warning light is *blue*, a bruise is *purple*. The word feels like a judgment made by the eye, and it is tempting to treat it as a coat of paint on the world: the object has the color, and we merely notice it. But for cells, color is not a label. It is a physical parameter of incoming energy, as crisp and measurable as the mass of a molecule. Light does not merely *show* tissue; it can *push* on it, heat it, and—if a protein is built to take advantage—kick it into a new shape.

The first step is to replace the painterly idea of color with a clock-

and-ruler idea. Light is an electromagnetic wave: it has a wavelength (the distance between wave peaks) and a frequency (how often those peaks pass a point each second). Wavelength and frequency are tied together by a simple constraint: in a given medium, longer wavelengths mean lower frequencies. A photon packages that wave into a discrete unit of energy, and the key point is that the energy of a photon is set by frequency. Higher frequency means higher energy per photon. Since blue light has a shorter wavelength than red light, it also has a higher frequency, and therefore each blue photon carries more energy than each red photon.

That statement can sound abstract, so it helps to anchor it to something familiar. Sunlight contains a spread of wavelengths. If you pass it through a prism, you can fan those wavelengths into a spectrum—violet and blue on one end, red on the other. The spectrum is not just a palette; it is a gradient of photon energies. At the blue end, each photon is more *expensive*: it can pay for chemical rearrangements that a red photon may not be able to afford. At the red end, each photon is cheaper but tends to be less readily absorbed by many biological molecules, meaning it can travel farther before being stopped.

Already the practical tension is visible. If you want a light-sensitive protein to do decisive work—open a pore, move an ion, yank a molecular lever—you care about energy per photon. But if you want that light to reach its target deep inside brain tissue, you care about how far photons of that color can travel. In living flesh, those two desires are often pulling against each other.

A photon's energy does not come as warmth by default. It depends on what captures it. If a photon is absorbed by water or by a broad, messy set of molecular vibrations, the energy tends to turn into heat, distributed as microscopic jiggling. If a photon is absorbed by a carefully tuned pigment—one that has a particular set of electron energy levels—it can be converted into a specific chemical event: an electron is lifted, a bond twists, a molecule flips into a new configuration. Biology's photoreceptors, including those that optogenetics borrows, are built around this second possibility. They are not passive solar panels that simply warm up. They are molecular machines that use photon energy to bias a physical change in a predictable direction.

The fact that a blue photon carries more energy does not mean blue light is automatically *strong* in a way that helps experiments. The word *strong* hides several different properties: energy per photon, total power delivered, how likely a photon is to be absorbed, and how far the beam stays concentrated rather than washing out into a glow. Tissue makes these properties diverge.

Consider what happens when a narrow beam of light enters brain tissue. The brain is mostly water, but it is not optically smooth like a clear glass of water. It is full of membranes, organelles, protein complexes, blood vessels, and the wrapped fatty insulation of axons. Each of these features slightly changes the local refractive index—how fast light travels through that little patch. When light crosses boundaries between regions with slightly different refractive index, it scatters: its direction is nudged. Do this once and the beam is merely blurred; do it thousands of times and a tight spot becomes a diffuse halo. Scattering does not require the light to be absorbed; it is a loss of direction, not necessarily a loss of photons.

Shorter wavelengths scatter more strongly in most biological tissues. The rough intuition is that small structures disrupt short waves more effectively than long ones, the way a pebble makes a pronounced ripple pattern in a small wave but barely registers in a long swell. The practical consequence is immediate: blue light spreads out quickly. That spreading lowers intensity at any given point, and it limits how far a concentrated stimulus can reach. Red light, with longer wavelength, tends to maintain direction better and can penetrate more deeply before scattering turns it into a fog.

Absorption is the other half of the story. Some molecules behave like selective photon traps. Hemoglobin, the oxygen-carrying molecule in blood, is a major absorber in the visible range, especially toward the blue-green part of the spectrum, with absorption changing depending on whether oxygen is bound. Melanin in skin is another strong absorber, which is why skin tone affects how much light gets through the surface. Water is comparatively transparent in much of the visible range but becomes increasingly absorbent as you move into the infrared. Lipids, proteins, and other tissue components each contribute their own smaller fingerprints.

Put scattering and absorption together and you get a simple reason why "color" becomes an engineering variable in biology. Blue photons may have more energy, but in tissue they are more readily scattered and often more strongly absorbed by pigmented molecules. Red photons may be less energetic, but they can slip through tissue with less deflection and, within a useful window, less absorption.

There is a famous, everyday example of this that most people have encountered without thinking of it as optics. Shine a bright white flashlight against your closed hand in a dark room. The glow that comes through the skin is red. Your hand is not emitting red light. The blood and tissue are acting as a filter: shorter wavelengths are preferentially absorbed and scattered, while longer red wavelengths are more likely to make it through. The same physics is at work when surgeons use red or near-infrared light for certain imaging tasks, or when a medical sensor tries to "see" through tissue rather than just off its surface.

A concrete case where color becomes a tool is pulse oximetry: the small clip placed on a fingertip in clinics and ambulances. It shines two colors of light through the finger and measures how much emerges on the other side. Oxygenated and deoxygenated hemoglobin absorb those colors differently, so the ratio of transmitted light can be converted into an estimate of blood oxygen saturation. The device does not care about color as a sensory experience; it treats color as a calibrated probe into tissue chemistry, chosen specifically because hemoglobin's absorption behaves differently at those wavelengths and because red/near-infrared light can traverse a finger without being totally extinguished. The method only works because *red* and *not quite red* are physically distinct levers.

Phototherapy offers another story, one that brings us closer to the logic that later shaped optogenetics. In the late nineteenth and early twentieth centuries, Niels Finsen became famous for treating lupus vulgaris (a form of cutaneous tuberculosis) with concentrated light. The exact biology was not fully understood at the time, but the treatment was built on a crucial empirical discovery: some wavelengths produce therapeutic effects more effectively than others. Finsen's arc-lamp apparatus filtered and focused light onto lesions, and the approach helped earn him the 1903 Nobel Prize in Physiology or Medicine. A century later, newborns with jaundice are treated with blue-green light that converts bilirubin in the skin into forms the infant can excrete more easily. In both cases, the color is not cosmetic. It is matched to molecular absorption properties so that photons can drive a specific chemical change.

These medical uses also illustrate a caution: if you pick the wrong wavelength or deliver too much power, light becomes damage rather than information. Photons can create heat, and heat can injure. They can also initiate photochemistry that produces reactive molecules capable of harming cells. Blue light, because it carries more energy per photon, is more likely to push molecules into unwanted chemical reactions if absorbed by the wrong targets. In everyday life, this is part of why ultraviolet light—shorter wavelength than visible blue—is so biologically aggressive: it can break chemical bonds and damage DNA. Visible blue is less extreme, but the direction of the risk is the same. Red light tends to be gentler per photon, but "gentler" is not a free pass; if you dump enough red light into tissue, the energy still becomes heat. Thermal injury is indifferent to the aesthetic of the illumination.

For experiments that aim to control cells inside the brain, all of these considerations collide. A neuron is an electrical cell, and changing its voltage requires moving charge across its membrane. Light cannot directly grab ions and drag them through lipid; the photon has to be caught by a molecule built to translate light into a mechanical or chemical action. That translation has to happen reliably, repeatedly, and on a timescale compatible with fast electrical signals. But the light must also be deliverable to the right place without cooking it, bleaching it, or scattering so widely that precise targeting is lost.

This is where it helps to think of photons almost as a kind of currency. You can pay your target protein with many low-value coins (lots of red photons) or fewer higher-value coins (fewer blue photons). But the path to the target is full of toll booths: scattering steals direction, absorption by blood steals photons entirely, and conversion to heat steals safety margin. The "price" you pay in photons must be high enough that the target protein receives sufficient absorption events to do its work, yet low enough that the surrounding tissue does not pay the cost.

The simplest way to increase the number of photons reaching a target is to increase light intensity. That strategy runs into two immediate limits in brain tissue. First, intensity falls rapidly with distance because of scattering and absorption. Second, intensity is not distributed evenly; hotspots form near the light source. A tiny optical fiber can deliver enough light to activate proteins a millimeter away, but it can also deliver far more than necessary to cells right next to the fiber tip. Those nearby cells are at the greatest risk of heating and phototoxicity, even though they are not necessarily the ones you most care about.

Color choices shift where those limits bite. Blue light often gives robust activation of many commonly used light-sensitive proteins, partly because those proteins absorb strongly in the blue range. Yet blue's tendency to scatter means the stimulated volume can be shallow and somewhat diffuse. Red-shifted light—longer wavelength, edging toward the near-infrared—can reach deeper tissue with less scattering, which can be the difference between stimulating superficial layers and reaching a structure buried beneath them. But red photons bring less energy each, and not every light-sensitive protein can use them efficiently. A protein tuned to blue may simply ignore red light; the photon arrives with the wrong "key" for its absorption lock.

Tissue optics also reshapes the meaning of "precision." A laser can be focused to a fine point in air. In brain tissue, scattering turns that point into a glow. A blue beam may start sharper but blur quickly; a red beam may retain focus longer. If your goal is to isolate a small group of cells, the scattering properties can matter as much as the genetic targeting. In other words, the optical physics can become the limiting factor in how "cell-specific" a manipulation feels in practice.

There is an additional subtlety: the brain is not an optically neutral medium across time. Blood flow changes with physiological state, and blood absorbs light. If a region has more blood volume at a given moment, it can become slightly darker to certain wavelengths. This does not usually dominate an experiment, but it is one more reason that light delivery is not simply a matter of setting a power knob. The living brain is dynamic, and its optical properties shift with that dynamism.

Safety constraints in laboratories end up echoing safety constraints in medicine. Engineers worry about heating because a few degrees

Celsius can change neural firing patterns and synaptic behavior even without causing obvious damage. Biologists worry about phototoxic chemistry because certain wavelengths can generate reactive oxygen species when absorbed by cellular components, especially under intense illumination. And everyone worries about the practicalities: a light source must be stable; an implanted fiber must not break; an animal must be able to move; the stimulation must be repeatable across days. In that practical world, "blue versus red" is not a philosophical question about the nature of color. It is a negotiation with tissue: how much intensity can be delivered, how far it will reach, how much it will spread, and what collateral effects will accompany it.

All of this might make light sound like a blunt tool—something that floods tissue rather than touches it. Yet biology has been using light with exquisite specificity for a very long time. Vision begins when a single photon is absorbed by a pigment molecule in a photoreceptor cell. Plants time their flowering and growth direction using light-sensitive proteins tuned to particular colors. Microbes harvest light to pump ions across membranes. Nature did not solve photoreception by brute force, blasting cells with illumination until something yields. It solved it by building proteins that treat photon absorption as a trigger: a tiny, reliable, quantized event that can be amplified into a cellular action.

The underlying promise of optogenetic tools rests on the same premise. If a protein can be made to respond to light in a predictable way, then color becomes a handle on cellular behavior. But that handle is only as good as the physical path between the light source and the protein. In brain tissue, that path is shaped by scattering, absorption, heating, and the energy-per-photon that comes bundled with wavelength. Color, in other words, carries force not because it looks vivid to us, but because it sets the terms of the bargain between photons and living matter.

3.2 The Molecule That Twitches in the Dark

A beam of light feels continuous to us, like water from a tap. At the scale that matters for photoreceptors, it arrives in grains. One photon is absorbed here, another there, and for long stretches nothing at all. In a dim room, a retinal cell in the eye may wait in literal darkness for a single photon to land in exactly the right place. That scarcity is not a bug; it is the central constraint. A photoreceptor cannot rely on warming up gradually, the way a black shirt warms in the sun. It needs a mechanism that turns one absorbed photon into a decisive internal event—small, yes, but unambiguous.

That event begins with a molecule so modest it is easy to underestimate. It is called retinal (pronounced *RET-in-all*), and it is derived from vitamin A. The name is not an accident: retinal was first understood through the physiology of the retina and through the clinical consequences of vitamin A deficiency. Long before anyone could see a protein structure, physicians knew that diets lacking vitamin A produced night blindness. Something in the eye needed a vitamin-derived ingredient to respond to faint light. Retinal is that ingredient, a light-catching chromophore—a term that simply means "color bearer," the part of a system that actually absorbs photons.

A chromophore is necessary because most proteins, by themselves, are not great at absorbing visible light in a specific and useful way. Proteins are built from amino acids, and amino acids tend to absorb strongly in the ultraviolet rather than the visible. If a protein needs to be triggered by blue-green light—the kind that travels well through water and is abundant in daylight—it often recruits a helper molecule with the right electronic properties. Retinal provides that helper function for opsins.

Opsins, as we saw, are the protein scaffolds that hold the light-sensitive machinery. Retinal is the spring-loaded latch inside that scaffold. It is not loosely associated, like a coin in a pocket; it is part of the protein's mechanical core. In vertebrate rhodopsin, retinal is covalently bound to opsin via a protonated Schiff base (classically Lys296 in vertebrate rhodopsin). "Covalent" means the bond is as chemically committed as it gets: the retinal is attached by shared electrons, not mere proximity. A "Schiff base" is a particular kind of link involving a nitrogen atom; "protonated" means that nitrogen carries an extra proton, and with it an electrical charge that helps tune retinal's light absorption. Lys296 is the amino acid lysine at position 296 in the protein chain, and it acts as the anchoring point. This single tether matters because it ensures that when retinal moves, the protein must feel it. The chromophore cannot simply absorb light and wiggle harmlessly; its motion is mechanically coupled to the opsin.

Retinal's job is not to fluoresce prettily or to store light energy for later. It performs a specific motion, almost like the snap of a mouse-trap: it changes shape.

To picture the shape change, imagine retinal as a small rod with a bend in it. In the dark, it sits in a kinked form called *cis*. After absorbing a photon, it straightens into a form called *trans*. Chemists call this a cis–trans isomerization. An isomer is a molecule with the same atoms but arranged differently; cis and trans are two such arrangements around a double bond, a region of the molecule that behaves like a stiff hinge. In the cis form, parts of retinal lie on the same side of that hinge; in the trans form, they lie on opposite sides, making the overall molecule straighter.

That sounds like a quiet adjustment, the sort of thing that might happen when a molecule is warmed. The startling part is the timescale. In rhodopsin, photoproduct formation within $\sim$ 200 femtoseconds has been experimentally observed. A femtosecond is 10^{-15} seconds. Light itself travels only about the width of a human hair in that time. The first mechanical twitch—the decisive snap from cis to trans—

starts essentially as soon as the photon is absorbed.

The speed forces a change in intuition. It is tempting to imagine a protein "noticing" light and then, after some thinking, responding. In reality, the earliest response is closer to a reflex built into physics. The photon's energy lifts retinal into an excited electronic state, and that excited state reshapes the landscape of forces within the molecule. The molecule takes the quickest downhill route available, and one of those downhill routes is a twist around the hinge. The protein does not wait to be instructed; it is designed so that the instruction is built into the molecule's permitted motions.

A vivid historical moment came when scientists learned how to watch this happen in real time. In the early 1990s, ultrafast "pump–probe" experiments made it possible to excite rhodopsin with one laser pulse (the pump) and then, a controlled fraction of a trillionth of a second later, interrogate its state with a second pulse (the probe). Schoenlein and colleagues (1991) used such methods to capture the earliest photoproduct after $\sim$ 500 nm excitation. The headline was not merely that rhodopsin changes quickly; it was that the primary photochemical event is over before many molecular vibrations can even share the energy out. The photon is not gently melted into heat. It is funneled into a directed change.

If retinal simply straightened in empty space, the story would be neat but biologically incomplete. The deeper trick is that retinal straightens while embedded in a tight protein pocket. The opsin does not offer a roomy dance floor; it offers a narrow corridor. The chromophore is constrained on purpose, held under tension like a compressed spring. When retinal flips, it does not relax into an ideal, comfortable trans form right away. Early intermediate bathorhodopsin contains a strained all-trans chromophore configuration. "Strained" is the key word. The chromophore is now in a shape that it wants to adjust further, but cannot without pushing on the surrounding protein. The protein has, in effect, trapped photon energy as mechanical stress.

That is the moment when "absorbing light" becomes "doing work." Absorption alone is an accounting event: energy enters the system. Work is directional: something changes in a way that can be harnessed. By forcing retinal to isomerize within a confined pocket, opsin converts the photon's energy into a shove against its own structure.

It helps to think about why evolution would build such a precarious arrangement. Photoreceptors must be quiet in the dark. A random thermal jiggle should not frequently trigger false signals, or the cell would hallucinate light. The dark state of rhodopsin is a carefully stabilized configuration: retinal in the 11-cis form (in vertebrate rods) held in place by the opsin. In darkness, it is metastable—stable enough to persist, but primed so that the right energy input can tip it. The photon provides that energy with precision. Then the isomerization, because it is mechanically coupled and produces strain, is very unlikely to be mimicked by ordinary thermal noise. The system is biased toward truthfulness: one photon, one twitch.

A second reason for the covalent tether becomes clear here. Many sensors in biology rely on loose binding and unbinding: a molecule docks, then later floats away. That can be slow and variable, and it can require the cell to keep a supply of the docking molecule around. By contrast, retinal is permanently attached to opsin. The system is always assembled, always ready, always in the same initial geometry. Reliability is a kind of speed.

The protein pocket also acts like a tuning fork for color. Retinal in isolation absorbs light differently than retinal in a particular opsin. By altering the electrical environment around the protonated Schiff base and by shaping the chromophore's geometry, opsins shift the wavelength at which retinal most efficiently absorbs. This is part of how different visual pigments in cones are tuned to different parts of the spectrum. It is also part of how microbial and algal opsins end up with action spectra that make sense in their ecological niches. The chromophore is the same basic molecule; the protein context changes the color it "prefers."

A third reason is that the strain can be released in a controlled sequence. Rhodopsin does not jump from dark to fully active in a single step, like a light switch snapping on. It passes through a series of intermediate states—names like bathorhodopsin and metarhodopsin mark those steps—each one a slightly different compromise between retinal's new shape and the protein's previous architecture. That cascade creates opportunities for timing and for regulation. Some steps are ultrafast; others are slower rearrangements as helices shift and the protein finds a new stable form. The earliest twitch is the spark; the later movements are the translation into a cellular signal.

The scale shift can be hard to hold in mind. Light entering the eye or an optical fiber seems large and macroscopic. Retinal is a molecule; opsin is a single protein; the critical motion is a twist around a bond. Yet that twist is not microscopic trivia. It is the trigger for electrical events that can change behavior.

A simplified cartoon can help fix the logic without drowning in chemistry: retinal is kinked in darkness, straightened by a photon, and that straightening pushes on the opsin, which sits in the membrane.

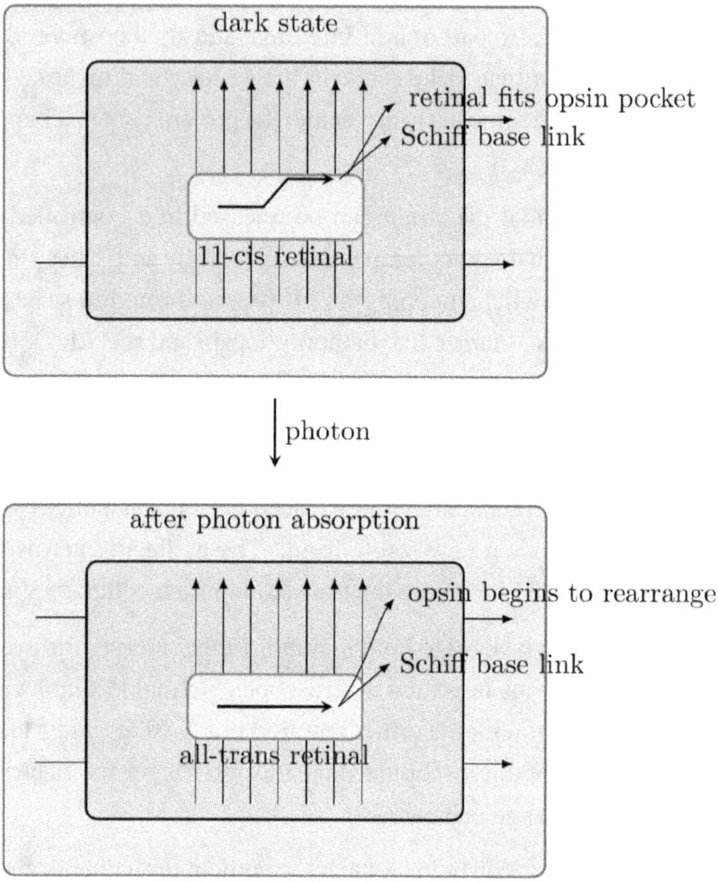

A diagram like this is deliberately crude, because the conceptual point is crude in the best sense: a photon triggers a shape flip, and the flip pushes on a membrane protein. The deep sophistication is hidden inside the words "shape flip," because that flip is a kind of mechanical logic gate—off in the dark, on after absorption.

The idea of a "twitch in the dark" also captures something emotionally human about photoreception. Vision feels effortless, but its molecular trigger is a high-wire act. A photoreceptor balances two incompatible goals: it must be stable enough not to fire randomly, and sensitive enough to respond to tiny amounts of light. Retinal's cis configuration, held under tension by opsin and pinned by a covalent bond, is a way to store potential action while staying quiet. The photon is the only acceptable excuse to move.

Retinal's origin in vitamin A adds another layer to the story. Vitamin A is not synthesized by the human body in sufficient quantities; it comes from diet, either directly from animal sources or indirectly from plant carotenoids that the body converts. The link between nutrition and a single molecular event—one photon causing one isomerization—helps explain why deficiency produces such a specific symptom profile. Night vision fails early not because the eye is generally weak, but because rods operate near the limit where individual photons matter. When the chromophore supply falters, the entire apparatus is left with proteins that cannot be triggered.

Retinal's covalent tether to opsin also hints at why photoreceptors are so exquisitely reproducible. In many signaling systems, variability creeps in because molecules must find each other by diffusion, bind for a while, and then unbind. Here, the key elements never part ways. The chromophore is not recruited anew every time; it is installed. That installation allows the cell to build large arrays of identical sensors—millions of rhodopsins in a rod outer segment—each one a small, resettable trap for a photon.

The reset is crucial, too, because isomerization is not a one-way ticket. Retinal can be returned to its original cis form through a biochemical cycle, allowing the photoreceptor to be used again. Nature's design separates the ultrafast trigger (the photoisomerization) from the slower housekeeping needed to restore the dark state. The fast part is physics, the slow part is metabolism, and the two are coupled only by the fact that the same molecule sits at the center of both.

There is a final subtlety worth holding onto because it prevents a common misunderstanding. Retinal's isomerization is not the "signal" by itself, in the sense that a neuron directly reads "cis" versus "trans" as a conscious sensation. It is the first irreversible-looking step in a chain that couples light to cellular actions. In vertebrate vision, that chain includes biochemical amplification: one activated pigment can influence many downstream molecules. In microbial photoreceptors—the ones that later became so useful for controlling neurons—the chain can be shorter, sometimes almost brutally direct: retinal twists, the protein opens a pathway for ions, electrical charge moves.

The essential unifying idea is that a single photon can be made to *matter*. It matters because retinal is a chromophore with the right absorption properties. It matters because retinal is covalently bound to opsin via a protonated Schiff base (classically Lys296 in vertebrate rhodopsin), ensuring the protein cannot ignore the motion. It matters because the primary twitch is ultrafast, with photoproduct formation within $\sim$ 200 femtoseconds observed in rhodopsin. It matters because the earliest product, bathorhodopsin, contains a strained all-trans chromophore configuration, turning photon energy into mechanical stress rather than letting it dissolve into warmth. That is how light, which seems so intangible in the hand, becomes a forceful, repeatable trigger inside living cells.

3.3 From Flash to Flow

Retinal's snap from cis to trans is a private event inside a protein pocket, too small to impress a neuron by itself. The membrane does not care that a chromophore has changed shape; it cares about charge. Neurons live and die by where ions sit—sodium, potassium, chloride, protons—on either side of a thin lipid sheet that behaves like an electrical capacitor. The only way light can reliably steer a neuron is by changing ion movement across that sheet. That is the conversion that matters: a photon becomes a change in protein shape; protein shape becomes ion flow; ion flow becomes a change in voltage; voltage shifts the odds that the cell will fire a spike.

Opsins solve the conversion problem in two main styles. One style opens a pathway and lets ions move according to forces that are already there. The other style acts more like a microscopic engine, using photon energy to push ions *against* those forces, building an electrochemical gradient the cell can cash in later. The first style is a light-gated channel; the second is a light-driven pump. They can be made from related protein parts, and they both begin with retinal, but they play distinct roles in biology and, later, in neurotechnology.

A useful way to feel the distinction is to imagine a dam. A channel is a floodgate: pull a lever and water rushes down the slope because gravity was waiting all along. A pump is a powered screw: it can lift water uphill, storing energy in the height difference. Both can move water, but only one can do it without paying an energy cost at the moment of movement. In cells, the *slope* is an electrochemical gradient: a combination of concentration difference and electrical attraction or repulsion.

A light-gated channel is, at heart, a pore through the membrane. In the dark it is closed; after photon absorption it changes conformation and becomes open enough for ions to pass. Once open, ions do not need to be shoved. They diffuse and drift through the pore because the gradients are already there. If the inside of a neuron is negatively charged relative to the outside, positively charged ions feel an electrical pull inward. If sodium is more concentrated outside than inside, it also feels a chemical pull inward. The channel does not supply the motive force; it supplies the doorway.

A light-driven pump is a stricter machine. It binds an ion on one side, changes conformation through a sequence of steps, and releases the ion on the other side, repeating this with each cycle of light-driven shape changes. In microbial proton pumps, this cycle is called a photocycle because the protein goes through a reproducible set of intermediate states after illumination. The key point is that the pump can be directional and can move ions in a way that builds gradients rather than merely letting them run down.

The historical roots of these ideas are older than optogenetics by decades and come from organisms that do not have nerves at all. In 1971, Dieter Oesterhelt and Walther Stoeckenius described a rhodopsin-like protein from the "purple membrane" of the halophilic microbe *Halobacterium halobium*. This was a surprise. Rhodopsin had been culturally tied to animal eyes, to seeing. Here was a microbial pigment protein, abundant enough to form visible patches of purple, doing something else with light. Within a couple of years, work by Oesterhelt and Stoeckenius connected that photochemistry to membrane function: light could drive proton transport, helping the cell store energy as a gradient.

That story matters because it clarifies what retinal-based proteins can be used for. In animal vision, the opsin does not directly pass ions. It triggers chemistry. In microbial rhodopsins like bacteriorhodopsin, the opsin directly moves ions, creating a proton motive force that can be converted to ATP. In some algal proteins, the opsin is itself an ion channel: light opens it, and current flows. The same basic chromophore trick—retinal isomerization—has been wired to radically different outputs.

Animal phototransduction is worth holding up as a contrasting design, because it shows one extreme: sensitivity through amplification. In rod cells, a single activated rhodopsin can trigger a measurable electrical response via enzymatic amplification: R^* activates multiple transducins; activated PDE hydrolyzes many cGMP molecules; cGMP channel closure changes membrane current. The letter R^* is shorthand for the activated form of rhodopsin, the state competent to engage its partner proteins. Transducin is a heterotrimeric G-protein; PDE is phosphodiesterase; cGMP is cyclic guanosine monophosphate. The names are less important than the logic. One photon changes one rhodopsin. That rhodopsin catalyzes changes in many transducins. Those changes, in turn, lead to the destruction of many cGMP molecules. That chemical change closes cGMP-gated channels and reduces inward current, causing the rod cell to hyperpolarize.

That last twist—light leading to hyperpolarization—can feel counterintuitive if one has absorbed the simplistic notion that stimulation always excites. Rods in darkness are relatively depolarized and continuously releasing neurotransmitter; light reduces that release. The key for the present discussion is that the opsin is not a pore. It is a chemical switch that starts a cascade, and the cascade is where the electrical change is generated.

Microbes often make the opposite trade. They may not need single-photon sensitivity. They need robust energy capture and simplicity: one protein doing one job, reliably, without an elaborate web of enzymes. In microbial proton pumps (bacteriorhodopsin), light-driven retinal isomerization launches a photocycle that transfers protons across the membrane, generating a proton motive force convertible to ATP. Here, the photon's energy is stored as a gradient: more protons on one side than the other. That gradient can later drive ATP synthase, the molecular turbine that uses proton flow to make ATP. Light, in other words, becomes metabolic currency.

Between these sits a form that turned out to be the optogenetic sweet spot: the light-gated channel. In channelrhodopsin-2, the *flash to flow* mapping is direct: light opens a cation conductance intrinsic to the single protein (not a second-messenger cascade), producing immediate ionic current. *Cation* means positively charged ions. A conductance is an ease of current flow; if you increase conductance to cations, positive charge can enter the cell more readily.

Why is this so potent for a neuron? Because neuronal spikes are threshold events. A neuron does not gradually decide to fire in a proportional way, like a dimmer switch. It integrates currents—synaptic currents, intrinsic currents, injected currents—and when the membrane voltage crosses a critical threshold, voltage-gated channels open rapidly and an action potential is produced. Anything that nudges the membrane voltage closer to threshold can raise the probability of a spike; anything that pushes it farther can suppress firing.

A light-gated cation channel, when opened, often allows net positive charge to move into the neuron. In many neurons at rest, the inside is negative enough that sodium and other cations are pulled inward. The inward current depolarizes the membrane, shifting voltage toward spike threshold. If the channel opens quickly and the light is timed in short pulses, the resulting voltage change can be fast enough to drive spikes with millisecond precision.

A light-driven pump can also change voltage, but in a more forceful and sometimes less subtle way. If a pump moves positive charge out of the cell, it tends to hyperpolarize: the inside becomes more negative. If it moves negative charge in, the effect can be similar. Hyperpolarization makes spikes less likely by moving the membrane away from threshold. Pumps are attractive when strong silencing is desired, because they can push voltage in a consistent direction, cycle after cycle, as long as light continues. The cost is that pumps move a limited number of ions per cycle, which can make the currents smaller than those from an open channel, and the net effect depends on how many pump molecules are expressed and how fast they cycle.

This channel-versus-pump contrast can be stated in one sentence, but it is easier to appreciate through an everyday analogy. Think of a crowded theater. A channel is a door that suddenly swings open: people stream through rapidly, driven by the pressure of the crowd. A pump is a turnstile that clicks one person at a time, reliably enforcing direction. Doors create big flows quickly but do not determine who "should" move; turnstiles impose direction but limit throughput. Biology uses both designs because it has both kinds of needs: sometimes a cell wants fast electrical signaling; sometimes it wants to build and maintain gradients with tight control.

A neuron, viewed electrically, is a living accounting system for charge. Currents enter and leave through channels and pumps, and the membrane voltage is the running total. A light-sensitive channel or pump adds a new term to that accounting system, but it does so in a way that can be cleaner than many drugs or electrodes. Drugs often act through receptors that trigger cascades, and those cascades can have many downstream effects. Electrodes inject current but do not discriminate between cell types; they excite whatever membrane is nearby. A microbial opsin expressed in a targeted population can add a relatively specific, light-controlled conductance or transport pathway right where it is needed.

The speed of the conversion is part of its appeal. Retinal isomerization is ultrafast; channel opening can follow quickly enough to reshape membrane voltage on the timescale relevant for spikes. That is not merely a technical luxury. Timing is one of the core languages of neural coding. Two neurons may fire the same number of spikes but at different times relative to a rhythm or a sensory event, and that timing can change behavior. A method that can deliver currents with millisecond control does more than raise or lower activity; it can edit the temporal pattern of firing.

At the same time, the membrane is not a blank canvas. The same light-gated channel can have different effects in different cell types or states because the driving forces depend on existing gradients and membrane voltage. Open a cation channel in a neuron sitting near threshold and you may get a spike. Open it in a neuron that is strongly inhibited by synaptic input and you may get only a small depolarization. Open an anion channel in a neuron whose chloride gradient is unusual, and the effect can flip sign. These nuances do not undermine the basic causal chain; they make it biological rather than idealized.

Light-driven pumps also bring biological baggage in a subtle way. Because pumps move ions directionally, prolonged activation can alter intracellular or extracellular ion concentrations. Move too many protons across a membrane and local pH can shift; move too many chloride ions and chloride gradients can drift. In many experimental settings these effects are manageable and small, but they are reminders that a membrane is not just a wire. It is an active interface whose gradients are maintained by metabolism.

The central chain remains simple enough to keep in mind at a dinner table. A photon arrives and is absorbed by retinal. Retinal changes shape and forces an opsin to rearrange. That rearrangement either opens a pore or drives an ion across the membrane. Ion movement is electric current. Current changes membrane voltage. Voltage changes the likelihood of spikes. The remarkable part is not that any one step is mysterious; it is that evolution produced proteins where the entire chain can be compressed into a single genetically encoded component.

That compression is also a kind of honesty. In animal vision, the magic is amplification: chemistry multiplies the effect of one photon until the nervous system can register it. In microbial pumps, the magic is energy storage: light becomes a reusable gradient. In light-gated channels like channelrhodopsin-2, the magic is immediacy: light becomes current with minimal bureaucracy. Once that is clear, it becomes much easier to understand why certain opsins feel like natural actuators for neurons, while others feel like elegant but indirect sensors.

At the end of the day, neurons do not "read light." They read voltage. Opsins make light legible to the membrane by turning illumination into charge movement, and charge movement into shifts in excitability. A flash becomes a flow, and a flow becomes the possibility of a spike.

3.4 Why Algae Built the Perfect Prototype

A neuron has the luxury of living in a body. It is fed, protected from the weather, bathed in salt water kept within narrow limits, and connected to a vast support system of glia, blood vessels, and organs that regulate temperature and chemistry. A single-celled alga gets none of that. It is a tiny bag of chemistry drifting through a world that can turn hostile in minutes. For an alga, light is not only information, and not only energy; it is a hazard and a compass at the same time.

Too little light and photosynthesis starves. Too much light and the cell's photosynthetic machinery can be damaged, generating reactive molecules that do the cellular equivalent of a power surge. Light also comes with direction. A pond is not uniformly illuminated: shadows from leaves, ripples, and the daily arc of the sun create gradients. In the open ocean, depth and turbidity make the light field a layered map. For a small organism that can swim, steering in response to light can be the most important decision it makes, repeated constantly: approach, retreat, turn, stop.

That is phototaxis—movement toward or away from light. It is easy to romanticize it as a primitive "sense," but it is more like a navigation algorithm forced into the shape of a single cell. Phototaxis has to be fast. Waiting seconds to decide is, for a microbe, an eternity: currents can sweep it away, predators can catch it, or it can drift into a damaging intensity band. Phototaxis has to be reliable. A random twitch is wasted energy; a wrong-way decision can be lethal. And phototaxis has to be compact. A cell with a small genome cannot afford a sprawling signaling bureaucracy.

Those constraints explain why green algae became such an unexpectedly generous source of optogenetic tools. Their light detectors were shaped by selection pressure that sounds eerily like an engineer's wish list: speed, sturdiness, and a direct handle on membrane voltage.

A particularly instructive organism is *Chlamydomonas reinhardtii*, a single-celled green alga that swims using two whiplike flagella. Under a microscope it can look almost comical—an oval body with two beating tails—but it carries a sophisticated navigation apparatus on its surface: an eyespot. The eyespot is not an eye in the animal sense. It is a specialized organelle-like fraction enriched in photoreception/signaling proteins including ChR1/ChR2, supporting the idea that algae evolved a compact, spatially organized light-sensing module. It contains pigment granules that act as a shading screen, meaning the cell can compare light intensity as it rotates while swimming. With each turn, the eyespot alternately faces toward and away from the light source, turning a spatial problem ("where is the light?") into a temporal one ("is it brighter now or a moment ago?"). That temporal signal can be used to adjust the flagella's beat pattern and steer.

Steering requires a fast intermediate variable—something that can change quickly and influence movement machinery. In animals, that intermediate variable is often a cascade of chemicals that eventually affects ion channels. In *Chlamydomonas*, a crucial part of the intermediate variable is electrical: the membrane potential. Change the membrane potential and you can rapidly alter calcium entry and other ion-dependent processes that control the flagella. For that, a light-sensitive protein that directly passes ions is almost perfect.

This is where channelrhodopsins enter the story. Channelrhodopsin-2 (from *Chlamydomonas reinhardtii*) is a single-component, directly light-gated cation channel—providing a genetically encodable actuator that converts light to membrane depolarization without added signaling proteins. "Single-component" is not a marketing phrase; it is an evolutionary triumph. In one gene, the alga encodes a protein that sits in the membrane, binds retinal, and opens a pore when illuminated. No auxiliary enzymes are required to translate light into current. Light becomes an inward cation current. Inward cation current tends to depolarize. Depolarization is an immediate way to influence the rest of the cell.

The ecological logic shows up in the protein's optical tuning. In *Chlamydomonas*, channelrhodopsins are tied to native phototaxis/photophobic behavior with action spectra around $\sim$ 500 nm, consistent with rhodopsin-like photochemistry and fast electrical responses. "Action spectrum" means the set of wavelengths that are most effective at driving the biological response. Around 500 nm sits in the blue-green part of the spectrum—abundant in sunlight, transmitted well through water, and energetic enough to drive retinal isomerization. The alga is not choosing "pretty" colors; it is tuned to the photons it can depend on in its habitat.

One can watch the outcome in real time. In classic microscope observations, a population of *Chlamydomonas* in a dish will accumulate on one side when light comes from that direction, like iron filings drawn to a magnet. Alter the light direction and the swarm shifts. Increase the light to an unpleasant intensity and the sign of the response can flip, with cells fleeing rather than approaching. It is a tiny migration, but it is a behavioral decision: a computation rendered as motion.

The human story of how these algal proteins became known outside photobiology has the feel of scientific cross-pollination at its best. Work associated with Peter Hegemann and Georg Nagel was central in identifying and characterizing channelrhodopsins from green algae. Their efforts belonged to a long tradition of studying how microorganisms sense their environment—work often done for its own sake, because these organisms are elegant and because their simplicity offers clean experiments. Yet the properties that make channelrhodopsins useful for algae are the same properties that later made them irresistible for neuroscientists.

That resonance can be made concrete by comparing design requirements.

A neuron-control tool should respond quickly, because spikes occur on millisecond scales. Channelrhodopsins open on timescales compatible with fast electrical events, because algae cannot afford sluggish steering.

A neuron-control tool should alter membrane voltage directly, because voltage is the currency of spiking. Channelrhodopsin-2 is a light-gated cation channel; opening it produces immediate ionic current intrinsic to the single protein, rather than relying on a second-messenger cascade.

A neuron-control tool should be genetically encodable, because modern neuroscience prefers to target specific cell types by using genetic promoters and viral delivery. Channelrhodopsins are proteins: genes that can be inserted and expressed.

A neuron-control tool should be durable and repeatable, because experiments may require thousands of trials. Algae swim all day; their photoreceptors must endure repeated illumination without falling apart.

Vertebrate rhodopsin, for all its elegance, is not built to be such a tool on its own. In rods, rhodopsin is the trigger of a biochemical amplification chain; to transplant that chain into a neuron, one would have to transplant much of the cascade machinery as well. That is not impossible in principle, but it is no longer "single-component," and the timing and gain would be harder to control. Channelrhodopsins arrive as a compact module: light in, current out.

This is why optogenetics could adapt ChR2 for millisecond-timescale control of mammalian neurons by genetic expression and fast light switching, leveraging the same direct light-to-ion-flow design that algae use for steering. The moment ChR2 was expressed in a mammalian neuron and illuminated, the neuron acquired a new sensory modality: sensitivity to a chosen wavelength of light, not as a vague influence but as a membrane conductance that could be toggled with pulses.

It is worth pausing on how odd that is from an evolutionary point of view. Neurons do not normally have retinal-binding light-gated pores. The mammalian brain is a dark organ; evolution did not stock it with phototaxis hardware. The fact that an algal channel can be expressed and work in mammalian membranes reflects a deep unity in cell biology. Membranes are membranes. Ion gradients are ion gradients. Retinal is a small molecule that animal cells can supply or can obtain through diet and metabolism. A protein that changes conformation in response to retinal isomerization can, in many contexts, do its job as long as it is placed in a lipid bilayer and given retinal to bind.

The eyespot adds an additional conceptual gift: it shows that cells can solve spatial problems by clever geometry. The eyespot's shading pigments create a directional light signal without requiring a lens, an image, or a nervous system. That is a reminder that sophisticated sensing does not require a brain; it requires a physical arrangement that turns environmental structure into an internal signal the cell can use. In optogenetic experiments, researchers often recreate a version of that idea by controlling not just the timing of light but its location and pattern. The cell's membrane is a canvas. Where the light hits and how it spreads, as discussed earlier, becomes part of the experiment's meaning.

Channelrhodopsins also carry a small warning label, and it comes directly from their biological heritage. Nature does not build categories as neatly as textbooks do. Although channelrhodopsin-2 is used as a light-gated channel, mechanistic nuance/controversy exists: ChR2 can show proton transport/pump-like behavior under some conditions (sometimes described as a "leaky proton pump"), which matters for interpreting physiology and side effects (pH shifts) in native and engineered contexts. For an alga, a bit of proton movement may be tolerable or even useful as part of its electrical response. In a neuron experiment, prolonged or intense illumination could, in principle, create small local shifts in pH or ionic composition. In practice, experimental designs often avoid extreme illumination regimes, and many variants of opsins have been engineered with altered kinetics and ion selectivity. The deeper point is that a borrowed protein arrives with a biography: it was shaped by the needs of an organism, not by the clean desires of an experimenter.

That biography can be read in the protein's behavior. Algal navigation is not merely "turn on" and "turn off"; it includes responses to sudden changes in light, to sustained brightness, and to different wavelengths. The cell must distinguish between a momentary glint and a sustained dangerous exposure. It must adjust as it rotates. It must avoid saturating its own sensors. Many of the quirks that later occupy optogeneticists—desensitization under continuous light, recovery in the dark, variable conductance over time—are echoes of those natural operating conditions.

A concrete way to appreciate the elegance is to imagine *Chlamydomonas* as a tiny boat with two oars. When light arrives from one side, the eyespot sees a rhythmic flicker as the boat spins. That flicker becomes an electrical signal. The electrical signal alters ion flows. Ion flows alter the flagellar beat. The beat changes the boat's heading. The cell ends up steering toward a light level that suits it. The entire loop is closed within one cell, without a synapse in sight.

Neuroscience later borrowed the actuator part of that loop. Instead of using the current to steer flagella, the current can nudge a neuron toward or away from spiking. The borrowed component is small, but it is decisive: a membrane protein whose conformational change opens a pathway for ions. It is hard to overstate the convenience of that simplicity. When something goes wrong in a complex signaling cascade, there are many places to look: the receptor, the G-protein, the enzyme, the second messenger, the channel, the buffering system. When something goes wrong in a single-component light-gated channel, there are fewer moving parts: expression level, retinal availability, light delivery, and the channel's own kinetics.

The algae did not set out to make neuroscientists' lives easier. They were solving a survival problem: how to turn the physics of light in water into a directional behavioral choice. Evolution rewarded proteins that were quick, repeatable, and tightly coupled to membrane voltage. Channelrhodopsin-2 is, in that sense, not a miracle of modern engineering but a fossilized solution to an ancient ecological constraint—a small molecular device for translating illumination into electrical force, built to run all day inside a single, swimming cell.

Chapter 4

The Algae That Rewired Neuroscience: Channelrhodopsin's Origin Story

The road to optogenetics did not begin in a neuroscience lab, but in the odd behavior of single-celled algae that could swim toward light as if they possessed a tiny sense organ. Following that clue leads to one of the most consequential scientific transformations of the century: the moment biologists realized that a microbial protein could give neurons a light switch precise enough to test causation in the brain.

4.1 Phototaxis and the Accidental Blueprint

A single green cell in a drop of pond water can make a decision that looks, from a human scale, almost deliberate. A narrow beam of light enters from one side. The cell tilts, swivels, then swims into the brighter region as if it had *meant* to go there. No nerves, no brain, no eyes—just a body the size of a dust mote, propelled by two whip-like flagella. That small turn toward illumination is easy to romanticize, but the deeper marvel is that it can be treated as an engineering problem: what pieces of biological hardware let a cell measure light direction and convert that information into an electrical command that steers its motors?

Earlier, phototaxis was already on the table as a real behavior with real constraints. What changed the story was not that algae chased light—people had watched that for a long time—but that a few labs began to treat the behavior as something you could take apart with the same seriousness as a radio or a lock. If the cell can turn in a beam, then somewhere on or near its surface there must be a light-sensitive component. If it can turn quickly and reliably, the signal can't be purely slow chemistry. If it can turn in the right direction, the cell must be comparing light intensity over time as it rotates, turning that comparison into a left-versus-right correction in the beating of its flagella.

Chlamydomonas reinhardtii, the celebrity of this story, is a particularly obliging organism for this kind of work. It is large enough to handle in a lab, yet simple enough that its "sense and respond" machinery is not buried under layers of tissue. It also comes with a conspicuous feature: the orange-red eyespot, offset to one side, like a tiny compass rose painted on the cell. The eyespot is not an eye, but it is not merely decoration either. It is a physical arrangement of pigment and membrane that gives the cell a directional cue: as the alga swims and rolls, light reaches its photoreceptive machinery in a rhythm—bright, dim, bright, dim—and that rhythm contains information about where the light is coming from.

A concrete image helps. Imagine the cell as a little boat with two oars at the front (its flagella), and imagine a "window" on one side (the eyespot region) that makes the boat alternately squint and see clearly as it spins. If the bright phase lasts longer than expected, the cell adjusts the oars so the next spin brings more of the bright phase into view. The logic is simple; the implementation is not. The implementation requires molecules that can do at least three jobs: capture photons, translate that into an electrical signal at the membrane, and couple that signal to a change in flagellar beating.

Curiosity-driven work has a habit of producing these "parts lists" without knowing who will need them later. In the late twentieth century, microbial sensory biology was full of such lists. Bacteria that swim toward nutrients; archaea that move away from harmful light; algae that orient to find photosynthetically useful wavelengths. Each system raised the same question in a different accent: what is the minimum set of molecular components that can sense light and control motility?

The eyespot made Chlamydomonas especially attractive because it offered a rare luxury in biology: a visible structure that seemed to sit directly between stimulus and behavior. For many years, researchers could point to it and say, with some confidence, "the light sensing happens *there*." That is not the same as knowing how it works, but it narrows the search. The eyespot became a stage on which the drama of signal transduction might be watched in real time.

One of the most revealing lines of attack was to measure electrical currents triggered by light. This sounds technical, but the basic idea is familiar. A membrane is like a battery separated by an insulating wall. If ions are allowed to cross that wall through channels, current flows. In neurons, this is the language of spikes. In algae, it is the language of steering commands. By the early 1990s, Peter Hegemann and colleagues were among those who managed to record light-regulated currents in Chlamydomonas (Harz and Hegemann, 1991). That accomplishment matters for two reasons. First, it turned an animal-like behavior into something you could plot as a trace on paper: light in, current out. Second, it made it harder to believe that the entire response was mediated only by slow biochemical messengers. A current that rises and falls promptly when the light turns on and off suggests a mechanism that is electrically intimate with the membrane.

At that time, the identity of the photoreceptor remained elusive. It was easy to guess it might be "some sort of rhodopsin"—a family name for light-sensitive proteins best known from vision—because rhodopsins absorb light using a small molecule called retinal. Retinal is a molecular trick: it changes shape when it absorbs a photon, and that shape change can push on the surrounding protein like a spring. But "some sort of rhodopsin" was not a satisfying answer. There are many rhodopsins, and biology often reuses the same chemical idea for different ends. A key practical question hovered over the whole field: is the photoreceptor a detector that triggers a downstream chain of events, or is it closer to a switch that can directly perturb membrane voltage?

Here contingency enters the story. Many people were looking at microbial rhodopsins for reasons that had nothing to do with neuroscience. Some cared about evolution: why do organisms as different as algae and archaea use retinal-based photochemistry at all? Some cared about biophysics: can a protein be understood as a device that converts light energy into ion movement? Others cared about ecology: how does a cell place itself in the light field of a pond, where too little light starves photosynthesis and too much light damages the machinery that captures it?

This is also where the story becomes, quietly, about method. A big behavioral question—how does the cell chase light?—was reduced to smaller, testable ones. Which wavelengths are most effective? How fast is the response? Which ions are involved? What happens in mutants that fail to orient correctly? These questions may sound incremental, but they function like triangulation. If the behavioral sensitivity peaks at certain wavelengths and the current peaks at the same wavelengths, the same pigment is probably involved. If the current changes when calcium concentrations change, calcium likely plays a role. Piece by piece, the behavior becomes a set of constraints that any molecular explanation must satisfy.

By the early 2000s, that triangulation began to pay off at the gene level. In 2002, work linked phototaxis responses in Chlamydomonas to two rhodopsin photoreceptors that seemed tuned to low- and high-intensity light (reported that year in the *Proceedings of the National Academy of Sciences*). The headline is not merely "two receptors." It is that a behavioral program could be tied to specific genes that could, in principle, be cloned, expressed, and studied in isolation. That step is often where biology turns into technology: once a function is attached to a gene, the function becomes portable. Genes can be moved.

But portability alone does not make a revolutionary tool. The crucial ingredient was hiding in the type of rhodopsin these algae used. Animal vision, the rhodopsin most people have heard of, relies on a cascade: light activates the receptor, the receptor activates a G protein, the G protein triggers enzymes, second messengers change concentrations, channels open indirectly, and the cell's electrical state changes after a chain reaction. That architecture is excellent for amplification and sensitivity. A single photon can have a noticeable effect. Yet the chain also introduces delays and complexity. It is like turning on a lamp by sending a messenger to find the electrician, who then rewires the circuit breaker before the bulb lights.

Microbes sometimes use a different design philosophy: fewer steps, more direct action. If a single protein could both sense light and carry ions, it would collapse the whole chain into one component. That possibility had been hinted at by microbial rhodopsins known to pump ions—proteins that use light energy to move ions across a membrane. Pumping is direct, but it is also slow in the sense that each protein moves ions one at a time, per light-driven cycle, like a person carrying buckets of water. For steering a cell, or later for driving neurons with millisecond control, a bucket brigade is not ideal.

The accidental blueprint for optogenetics lay in the possibility of something even better: a light-gated *channel.* A channel is more like opening a floodgate: ions flow rapidly down their electrochemical gradient. If light could open such a gate, the membrane voltage could change quickly and robustly. In 2002 and 2003, Georg Nagel and Peter Hegemann, working with collaborators, identified and characterized channelrhodopsins from green algae as directly light-gated ion channels. Channelrhodopsin-1 was reported in 2002, and channelrhodopsin-2 in 2003 as a directly light-gated, cation-selective membrane channel. The names sound forbidding, but the underlying idea is startlingly simple: shine light, and ions pass.

To appreciate why this mattered, it helps to remember what a small, free-swimming alga is up against. Light in nature is not a gentle, constant spotlight. It flickers as water ripples, shifts as clouds move, and changes as the cell swims through gradients. A useful steering system must be both quick and resettable. If the cell turns toward light and then overshoots, it must correct itself within a fraction of a second. A receptor that triggers a long biochemical cascade might still work, but it risks sluggishness or saturation. A single-component, light-gated current source is tailor-made for rapid feedback.

The historical anecdote that captures the spirit of this research is less about a single eureka moment and more about the doggedness required to connect behavior to molecule. Measuring light-induced currents in a tiny, moving, light-sensitive cell is an exercise in experimental patience. The cell wants to swim away; the light used to probe it is also the stimulus; and the currents are small. Yet those early electrophysiological recordings provided a kind of "electrical fingerprint" of the photoreceptor system: turn the light on, see a current;

turn it off, see it relax; change the wavelength, see the sensitivity shift. That fingerprint gave molecular hunters something to match when candidate genes were finally isolated. When channelrhodopsins were expressed and produced currents with the right timing and spectral properties, it was not just a neat trick. It was a match between a decades-old physiological signature and a specific protein.

Channelrhodopsins also solved a conceptual problem that had lingered in phototaxis research: how to connect a photon to a membrane voltage change without inventing extra intermediates. In the language of mechanism, the receptor itself is the actuator. That short-circuited a lot of speculation. If the receptor is a channel, you do not need to posit a whole chain of second messengers to explain fast membrane currents. You can draw a straight line from light absorption to ion flow.

The eyespot now looked even more consequential. It was not merely a passive shading device but part of a spatially organized system in which a membrane region containing these proteins could generate localized currents that bias the behavior of the flagella. The cell did not need a nervous system to "compute" direction; it needed only an asymmetric sensor and a motor system that responded to the resulting electrical asymmetry. Phototaxis became a story of geometry plus membrane biophysics.

The broader microbial world provided further reinforcement that nature had already explored this design space. Retinal-based photoreception is not confined to algae; it appears in diverse microbes separated by vast evolutionary distances. That diversity is not a detail for taxonomists. It is a hint that microbial opsins are a versatile family of light-driven devices—pumps, sensors, channels—that evolution has tuned for different environments. Once you see them that way, you start to notice what an engineer would notice: modularity. A protein that converts light into ion flow is a module. It can, in principle, be inserted into any membrane and will do its job as long as the chromophore retinal is available.

For an alga, retinal is part of its native toolkit. For other organisms, the story varies, but the conceptual leap had already been made: the molecular machine is self-contained. That is what "single-component phototransduction" really means in practice. It is not that the cell does nothing else; it is that the core conversion from photon to current is handled by one protein rather than a whole relay team.

Here the "accidental blueprint" becomes visible. The blueprint was not a plan drawn by neuroscientists, but a set of properties that, in hindsight, reads like a wish list for controlling excitable cells. First, speed: light causes current quickly. Second, specificity: the protein responds to particular wavelengths. Third, genetic encodability: the machinery is written in DNA. Fourth, sufficiency: express the gene, provide retinal, shine light, and you have a controllable conductance. Those features are exactly what later made channelrhodopsins irresistible to people who wanted to intervene in neural circuits with precision, even though that was not the algae biologists' goal.

A useful way to feel the contingency is to imagine an alternate history in which algae used a purely indirect cascade like animal photoreceptors. Optogenetics might still have emerged, but it would have been clumsier. The control signals would have been slower, more variable, and harder to interpret. The fact that channelrhodopsins were channels meant that the timescale of light control could be pushed toward the timescale of electrical excitability. That alignment is the quiet hinge on which the whole field later turned.

The other ingredient, less glamorous but equally important, was the culture of basic science that kept the question alive. Phototaxis might have looked like an esoteric corner of microbiology, the sort of phenomenon that makes for charming microscope videos but not, at first glance, for medical breakthroughs. Yet the researchers who pursued it did so with the seriousness that mechanism demands. They cared about which currents flowed, which genes encoded the photoreceptors, and how these pieces were arranged in the cell. They were mapping a machine that nature had already built.

That machine had, in its own environment, a clear purpose. A Chlamydomonas cell lives in a world where light is both food and threat. It must position itself to harvest energy while avoiding photodamage. Its eyespot and photoreceptors are survival tools, not scientific curiosities. But survival tools can become laboratory tools when their operating principles are general. Channelrhodopsin was general in a way that was easy to miss until someone tried to transplant it: it spoke the language of membranes and ions, a language shared by algae, neurons, muscles, and many other cells.

A final, almost philosophical point is hiding in this origin story. Neuroscience often feels special because the brain is special. Yet some of the most transformative methods in brain science have arrived from places that seem, at first, far away from thought and feeling: bacterial enzymes that copy DNA, jellyfish proteins that glow, algae proteins that steer a cell toward the sun. The underlying reason is that life reuses a small set of physical principles. Light can flip a molecule; a molecule can change a protein's shape; a protein can allow ions to cross a membrane; ions can change voltage; voltage can trigger a spike or alter movement. The chain is continuous from pond water to cortex.

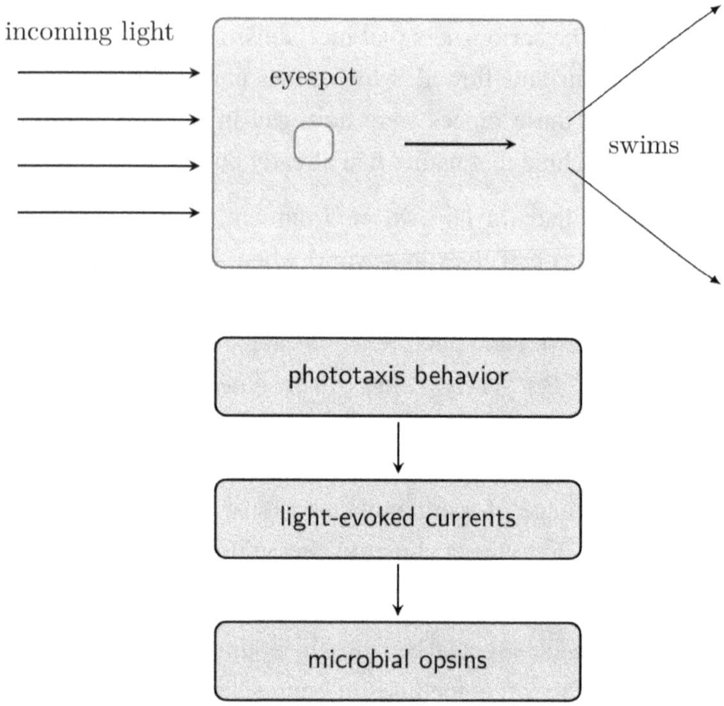

By the time channelrhodopsins were in hand as genes and as proteins, the original question about a cell's orientation in a beam had already done its most important work. It had forced a search for a molecular device that turns light into electricity with minimal delay, and it had revealed that such devices exist in nature not as curiosities but as solutions to everyday problems faced by single cells trying to live in a fluctuating world.

4.2 A Channel, Not a Cascade

Light is an oddly demanding stimulus for biology. It arrives in packets, at high speed, in particular wavelengths, and it can change instantly when a shadow passes or a cloud moves. Many living systems meet that demand by building elaborate molecular relay races. A photon is caught by a receptor; the receptor nudges a partner protein; that partner triggers an enzyme; the enzyme changes the concentration of a "second messenger" molecule; the second messenger recruits or modifies ion channels; only then does the cell's electrical state shift enough to influence movement, secretion, or gene expression. The logic is sensible: each step can amplify the signal and add opportunities for regulation. The cost is delay and biochemical clutter.

Channelrhodopsin broke that pattern in the most literal way. Instead of light being the first domino in a long chain, light was the hand on the doorknob. When the protein absorbed a photon, it altered its own shape so that ions could pass through it directly. No intermediary enzymes needed to wake up; no second messenger molecules had to accumulate; no distant channels had to be persuaded. The receptor was the conduit.

That sounds like a small mechanistic footnote until you think about the timescale of the brain. Neurons do not merely *fire* or *not fire*; they often convey information in the precise timing of spikes, separated by only a few thousandths of a second. In that regime, a control signal that arrives tens or hundreds of milliseconds late is not just sluggish—it can be a different message altogether. A tool meant to steer neural activity has to speak the nervous system's native dialect: fast, repeatable, and tightly locked to time.

The decisive insight, then, was not simply that algae possess photoreceptors. It was that one of their photoreceptors behaves like a light-driven electrical component: a single protein that converts photons into membrane current with startling directness.

A concrete way to grasp the contrast is to imagine two household devices. One is a motion-activated light that turns on only after a chain of events: the sensor detects movement, a small computer confirms the pattern, a relay clicks, the circuit closes. The other is a flashlight with a mechanical switch. Both can produce light, but one is built for interpretation and filtering, the other for immediacy. In biology, the "sensor-computer-relay" architecture is the cascade, and the "mechanical switch" architecture is channelrhodopsin.

The cascade style is deeply familiar from animal vision, and earlier discussions have already put the classic version on the table. What matters here is the general principle: cascades are powerful when you need sensitivity and signal processing, but they are awkward when you want a clean, time-stamped electrical perturbation. A cascade can also vary from trial to trial, because every step depends on concentrations, enzymatic states, and cellular context. That variability is not a flaw in nature; it is the price of a richly regulated system.

Microbial opsins live in a different design universe. Reviews emphasized that microbial opsins are functionally distinct from vertebrate visual rhodopsins and, at the level of primary sequence, unrelated (as summarized by Deisseroth and others). That detail can sound like biologists arguing about family trees, but it is actually a warning label for would-be tool builders: this was not a minor tweak to the molecules of eyesight. It was a separate evolutionary solution to the problem of light sensing, one that happened to be far more convenient for rapid electrical control.

Nagel, Hegemann, and colleagues put a name to that solution. In 2002, Channelrhodopsin-1 was characterized as a light-gated ion channel from green algae, reported as a proton channel. In 2003, Channelrhodopsin-2 was established as a directly light-gated, cation-selective membrane channel. Those two papers did more than add entries to a list of photoreceptors. They introduced a category: an opsin that *is* an ion channel.

For readers who do not think in ions, the phrase "cation-selective" can be translated as "it tends to pass positively charged ions." In an excitable membrane, allowing positive charge to enter generally nudges the voltage toward the threshold for spiking. The broad point is that this is not an indirect command delivered by chemistry; it is current, delivered to the membrane in real time.

The experimental work that established direct gating had a particular flavor: stripping a phenomenon down to its minimum moving parts. If you can express the candidate protein in a cell and then observe light-evoked currents without having to supply a whole algal signaling apparatus, the simplest interpretation is that the protein itself is responsible for the current. The appeal of such a "single-component" actuator is obvious in hindsight, but at the time it was an unexpected answer to the problem of phototaxis. The ecological question had been "how does a cell steer toward light?" The molecular answer was "by embedding a light-controlled electrical valve in its membrane."

Once you have that contrast in mind, several consequences fall out naturally.

First, direct gating makes light control *predictable*. If a protein's conformation is changed by a photon and that change creates a conductive pathway, then the main uncertainties are physical ones: how much light arrives at the membrane, how many channels are present, and what the ionic gradients are. Those are variables an experimenter can often estimate and manipulate. Cascades, by contrast, contain many hidden state variables inside the cell: enzyme activity, depletion of signaling molecules, feedback loops that adapt the response. These features are valuable for an organism's natural behavior, but they complicate attempts to use light as a clean input signal.

Second, direct gating makes light control *fast in a way that stacks*. A single flash can provoke a rapid current, but so can a train of flashes at regular intervals. That property matters when the goal is not merely to push a cell into a new steady state, but to drive patterns of activity that mirror the temporal structure of the nervous system. If you want spikes at 10 per second, or 40 per second, the control signal must reliably rise and fall on that schedule. A cascade that lingers and adapts can smear the timing, like trying to play staccato notes on an organ with a sticky key.

Third, direct gating makes the molecule *portable*. This is where the story begins to look like an accidental blueprint rather than a quirky detail of algal biology. A multi-protein cascade is hard to transplant: you need the receptor, plus its partner proteins, plus enzymes, plus the appropriate ion channels, plus the right cellular scaffolding to keep them near each other. A single-component channel is a self-contained module. Give a cell the gene, and the cell can build the protein and insert it into its membrane. Supply retinal (often already present in many animals' diets and tissues), and the protein can function. The point is not that no other biology matters, but that the core photon-to-current conversion is unusually self-sufficient.

That portability also helps explain why microbial opsins were so attractive once neuroscientists noticed them. Neurons are, in one sense, the most demanding clients imaginable: they are sensitive to perturbation, they encode information in timing, and they are embedded in circuits where unintended side effects propagate. A direct, light-driven conductance offered a degree of experimental cleanliness that other interventions struggled to match. Electrical stimulation is fast but indiscriminate; drugs can be specific but often slow and diffuse. A genetically specified, light-driven channel sat in an unusual sweet spot.

The story also has a human-scale lesson about how scientific surprises happen. The "channel, not a cascade" idea sounds like the kind of thing that would be obvious if you could watch the protein at work, but in real time it was not obvious at all. Photobiology had long been shaped by the success of cascade thinking, especially because cascades can produce dramatic amplification. A protein that simply *becomes* a pore when illuminated seems almost too straightforward, like discovering that a complicated machine is actually a lever. It takes a certain kind of experimental stubbornness to accept that nature sometimes chooses the lever.

One can glimpse that stubbornness in the way algal phototaxis research had been constrained by electrophysiological facts. Light could evoke currents rapidly. Those currents had specific spectral sensitivities, consistent with rhodopsin-like pigments. If the current is genuinely fast, then either a cascade is extraordinarily streamlined, or the receptor is intimately tied to ion flow. Channelrhodopsin offered a solution that fit the constraints with minimal extra machinery.

A separate subtlety, also highlighted in later reviews, is that microbial opsins and vertebrate rhodopsins are not simply the same protein used in two contexts. That matters because it helped free researchers from the expectation that light responses must follow the vertebrate playbook. In a sense, channelrhodopsin expanded the imagination of what an opsin could be. "Opsin" stopped meaning "a receptor that launches a cascade" and became "a light-driven machine," with multiple mechanical implementations.

A final way to appreciate why milliseconds loom so large is to consider how neuroscientists ask causal questions. Suppose you want to know whether a particular group of neurons can drive a behavior, or whether its activity is sufficient to bias a choice. If your intervention takes a second to build up and another second to decay, you may still change behavior, but you have blurred the link between neural activity and the animal's decisions. The brain's computations are not neatly boxed into one-second windows. They unfold rapidly, and they interact with ongoing oscillations, sensory inputs, and internal states. An intervention that can be aligned with a brief sensory cue or a moment of recall has a different kind of evidential power. Directly light-gated channels made that alignment plausible.

This is also why the notion of "single-component" is more than a biochemical curiosity. It is a promise about experimental control. When the same light pulse produces the same membrane current pattern, again and again, the experimenter can begin to treat the pulse as an input whose timing is known with confidence. That confidence, in turn, allows sharper hypotheses about circuitry: not merely that a manipulation changed something, but that a precisely timed perturbation was sufficient to alter a computation at a particular moment.

None of this diminishes the value of cascades in biology. Cascades are exquisitely suited for tasks like adapting to background light, integrating signals over time, or detecting extremely weak stimuli. But those virtues are not the virtues of a light switch. A switch is valued for being boring: it does the same thing each time, on the timescale you expect, without negotiating with the rest of the house.

Channelrhodopsin, once understood as a channel and not a relay, turned photons into something experimenters could schedule. Light stopped being merely a stimulus and became a trigger: a way to deliver a defined electrical nudge to a membrane on command. In a world where the nervous system's basic currency is time, that directness was not a minor convenience; it was the whole point.

4.3 The First Time a Neuron Obeyed Light

A neuron in a dish does not look like a thinking cell. Under a microscope it can resemble a spindly starburst, a small body with branching processes that wander across glass. Yet it behaves like a poised electrical device. With the right nudge, it produces a spike: a brief, stereotyped surge of voltage that travels along its membrane and, in networks, helps coordinate perception, movement, and memory. For more than a century neuroscientists have been able to *listen* to those spikes by placing electrodes near or inside neurons. The dream was to speak back in the same precise language, not with a blunt stimulus to everything nearby, but with a message aimed at a chosen population.

Channelrhodopsin-2 made that dream concrete once it was moved out of algae and into neurons. The conceptual leap was deceptively short: if ChR2 is a directly light-gated, cation-selective membrane channel, then in principle it should depolarize any membrane that expresses it, including a neuron's. The practical leap was enormous. A microbial protein had to be produced in mammalian cells, correctly inserted into the neuronal membrane, and activated with flashes of light that were fast, controllable, and gentle enough not to cook the tissue. And then, crucially, the output had to be clean: spike on cue, not "a bit more excitable," not a vague change that might or might not translate into firing.

The pivotal demonstrations arrived in 2005. Edward Boyden, Feng Zhang, and colleagues reported millisecond-timescale, genetically targeted optical control of neural activity using ChR2. The emphasis in that sentence is worth lingering on. "Millisecond-timescale" was a promise about timing. "Genetically targeted" meant the light sensitivity was conferred only on neurons that received the gene. "Optical control" meant the stimulus could be delivered without metal electrodes touching the cell. Together, those features implied something neuroscience had not previously enjoyed: an input that could be both selective and precisely timed.

Around the same time, a parallel 2005 study showed rapid optical activation of hippocampal neurons and network effects using green algae channelrhodopsin approaches. That parallelism was not just a race for priority; it was an early sign that the effect was not a laboratory curiosity that worked once under idiosyncratic conditions. When multiple groups can make a challenging technique work at roughly the same moment, it often means the components have finally become compatible: the protein is robust enough, the gene delivery is reliable enough, and the measurement tools are good enough to make the phenomenon repeatable.

To understand why the first neuronal spikes triggered by light felt like a threshold, it helps to picture the experiment as a conversation between three kinds of technology.

One is molecular biology: the ability to deliver genetic instructions into neurons. The phrase "viral gene delivery" can sound alarming, but in the laboratory it usually refers to viruses engineered to be harmless carriers. They are stripped of the machinery that makes them cause disease and repurposed as delivery vehicles for a chosen gene. Neurons are difficult cells to transfect by brute force; viruses have evolved to enter cells efficiently. By borrowing that trick, neuroscientists gained a way to persuade neurons to manufacture ChR2.

The second is optics and electronics: the ability to produce light pulses on command. A flashlight is not enough. The experiment demands a light source that can switch on and off quickly and reproducibly, and that can be aimed at a small region. In the early days, these setups often looked improvised: a microscope, a light path patched together with filters, and fast shutters or LEDs controlled by computer signals. What mattered was that the timing of illumination could be made as crisp as the timing of an electrical stimulus.

The third is electrophysiology: the ability to record spikes and currents in real time. Without that, a claim of "light control" could collapse into interpretive ambiguity. A neuron might change its firing rate because the light warmed the dish, or because the neuron's own light-sensitive molecules were perturbed, or because the stimulation was indirectly activating other cells. Recording allowed researchers to anchor the story in a trace: light pulse, membrane current, spike—or no spike—with latencies that could be measured in milliseconds.

The experimental logic was elegantly direct. Express ChR2 in a neuron, give a brief blue light pulse, and see whether the neuron fires. If the timing of firing locks tightly to the timing of light pulses over repeated trials, the result is not merely that light can influence a neuron. It is that light can serve as a reliable command signal.

The idea of "millisecond precision" can sound like marketing language until you translate it into laboratory practice. A typical spike is over in a couple of milliseconds. A neuron's decision to spike depends on how quickly depolarizing inputs arrive and whether they coincide with other inputs. If a tool can deliver depolarization with a sloppy onset, spreading over tens of milliseconds, it blurs into the brain's background chatter. If it can deliver depolarization sharply, it can be aligned to a sensory cue, a decision point, or a specific phase of an oscillation. Precision means that when you repeat the same pattern of light pulses, you get the same pattern of spiking with minimal jitter: the spike occurs at almost the same delay after the flash each time, not sometimes early, sometimes late, sometimes missing. That is the difference between "encouraging" a neuron and "instructing" it.

For many readers, it is hard to feel why a few milliseconds are so precious. A useful analogy comes from music rather than machinery. In an orchestra, the same notes played a fraction of a second late are not a small error; they are a different performance. Timing creates meaning. Neurons, too, can be sensitive to coincidence. A spike that arrives at one synapse just before another may strengthen a connection; arriving just after may weaken it. A tool that cannot control timing cannot easily ask questions about such rules.

Channelrhodopsin-2 was particularly well suited to this because its fundamental mechanism is so spare. As established in its foundational biophysical characterization, it is a directly light-gated cation channel. When illuminated, it produces a conductance that tends to pull the membrane toward depolarization. That tight coupling between photons and ionic current avoids the variable delays of cascades and the unpredictable diffusion of second messengers. The neuron's membrane becomes, in a sense, a screen on which light pulses can be written as electrical events.

Getting a microbial protein to behave properly in a mammalian neuron was not guaranteed. Proteins must fold into the right shape; they must be trafficked to the membrane rather than trapped inside the cell; they must not poison the host cell; they must find their chromophore. The success of ChR2 in neurons suggested that evolution had already delivered a remarkably portable device. ChR2 did not need an algal eyespot or algal signaling partners to do its job. The neuron provided the relevant physics—a membrane, ionic gradients, and an excitable threshold—and ChR2 supplied the light-controlled conductance.

A small historical detail captures how striking this portability felt. Algal phototaxis biologists had spent years connecting behavior to photocurrents in their organism of choice. Yet the moment that turned heads in neuroscience was not an algae turning toward a lamp. It was a mammalian neuron sitting still, impaled by a recording electrode, and firing a spike each time blue light flickered. The neuron did not "understand" light in any ordinary sensory sense; it had been endowed with a new input channel, a synthetic sensory modality. The cell was, in effect, bilingual: it still responded to synaptic currents from other neurons, but it now also responded to photons.

That bilingualism brought an immediate methodological transformation: it separated *where* you deliver a stimulus from *which neurons* receive it. Electrical stimulation, the older workhorse, ties those together. If you pass current through an electrode in brain tissue, you stimulate whatever excitable elements lie near the tip: axons passing by, dendrites, cell bodies of multiple neuronal classes, even fibers coming from distant brain regions. The stimulation is spatially localized but biologically indiscriminate. Drugs reverse the problem. A drug can target particular receptor types, but it spreads, and it acts on all neurons expressing those receptors, often over minutes. Optogenetic actuation promised a different pairing: biological selectivity through gene expression, and temporal control through light delivery.

The first demonstrations also did something subtler: they shifted how neuroscientists could argue about causation. Many older interventions were excellent at showing correlation or necessity in a broad sense. Lesion an area and the animal can't perform a task; block a receptor and a behavior changes. Those are powerful results, but they often leave open the question of sufficiency: is activity in a specific neuronal population, at a specific time, enough to produce a defined effect?

Sufficiency is a high bar, and it forces a particular kind of experimental honesty. If you want to claim that a particular set of neurons can drive an action, you need a way to activate those neurons in a controlled fashion, without simultaneously pushing the rest of the brain around. A light-controlled channel makes it possible to "write in" activity where it does not naturally occur at that moment and see what the system does with it. When a precisely timed flash elicits spikes in the targeted population, the experimenter can treat those spikes as an imposed cause rather than an observed correlate.

That does not mean the interpretation is trivial. Even with optogenetic control, the brain can respond in complex ways. Activating one population can recruit inhibition, trigger network oscillations, or produce compensatory activity elsewhere. Yet the causal claim becomes sharper: the induced spiking is not merely associated with the outcome; it precedes it with a known latency and is confined, as much as biology allows, to neurons carrying the gene.

The early 2005 work also highlighted a practical requirement that later histories emphasize: optogenetics is not one invention but a convergence. An actuator like ChR2 is necessary, but so is the ability to express it in targeted neurons and to deliver patterned light at behaviorally relevant timescales. A brilliant protein without gene delivery remains a curiosity; gene delivery without an actuator remains a method looking for a function. The success of the first neuronal experiments depended on both being ready at once, plus the mundane engineering of optical hardware that could be synchronized with electrophysiology.

One of the most telling aspects of those first light-evoked spikes is that they made a promise about reproducibility. Neuroscience is full of phenomena that can be elicited once, in one preparation, under delicate conditions. Such results can be scientifically valuable, but they do not immediately become communal tools. A communal tool is something you can teach to a graduate student, send to another lab, and expect to work again. The existence of parallel 2005 demonstrations hinted that ChR2-based photostimulation was crossing that threshold. It was becoming something one could build a research program around.

There is also an emotional component that is easy to overlook. For the scientists watching those traces, the moment carried a particular kind of thrill: a new form of command over a cell that had previously been accessible mainly through electrodes and pharmacology. A neuron responding to light can feel uncanny, not because it is mystical, but because it exposes how contingent the inputs to a brain cell really are. Neurons do not have a sacred relationship to synapses; they have a relationship to voltage and ions. If you can provide a new route for ions to move, you can provide a new route for influence.

That realization does not reduce the brain to a simple machine. If anything, it raises the stakes. When a tool can trigger spikes with crisp timing, it becomes possible to ask not only whether a group of neurons participates in a function, but whether a particular pattern of spiking is enough to bias perception, to retrieve a memory, or to tilt a choice under temptation. The flash of blue light is not interesting because it is blue. It is interesting because it can be treated as a programmable input to the very currency the brain uses: brief, timed electrical events.

In the end, the most radical feature of that first obedient neuron was not the spectacle of light controlling life. It was the new experimental grammar it enabled: instead of merely observing neural activity as it accompanied behavior, one could impose a precisely timed pattern on a chosen neuronal population and watch the rest of the circuit respond, spike by spike.

4.4 When a Tool Becomes a Movement

A striking laboratory result can be like a magic trick: dazzling the first time, hard to reproduce the second, and fragile in the hands of anyone who did not build the apparatus. A scientific movement begins when the trick becomes a craft. The craft has manuals, spare parts, calibration habits, and a shared sense of what counts as a clean control. It can be taught to newcomers, trusted by skeptics, and modified without breaking its logic.

Optogenetics crossed that threshold not because one experiment was beautiful, but because a whole toolchain solidified around it. Methods-focused histories have emphasized that the platform rests on a convergence of three components: microbial opsins, strategies for targeting expression, and optical interfaces for delivering patterned light. Each component is necessary and each is insufficient by itself. Together they created something scalable: a way to perturb selected neurons at selected times with an input that could be repeated across animals, labs, and years.

The early years after the first neuronal demonstrations in 2005 were full of the unglamorous work that determines whether a technique will stay local or go global. Consider what a lab actually needs in order to make neurons respond to light in a meaningful behavioral experiment. The opsin gene must be packaged so that it enters neurons efficiently. It must be placed under a promoter that yields robust expression without killing the cells or silencing the construct over time. The opsin protein must reach the membrane rather than accumulating in cellular compartments where it cannot conduct ions. Light must be delivered to the right brain region at intensities high enough to recruit opsins but low enough to avoid heating and photodamage. The stimulation must be timed with the experiment, whether that timing is locked to a sensory cue, a learned task, or spontaneous behavior. Finally, there must be a way to verify that the intended neurons were actually illuminated and that they actually responded.

That list is not glamorous, but it is the reason movements require more than ideas. Movements require infrastructure.

One concrete marker of infrastructure is the appearance of detailed in vivo protocols. A protocol paper is a kind of social contract. It says: *this can be done reliably, and here is the recipe*. It also reveals what had been implicit in early demonstrations: the method is not just a protein, but a bundle of practical choices. Which viral vector should be used for which brain region? How long after injection should one wait for expression to stabilize? What light powers are typical at the fiber tip? How should artifacts be checked—heating, light leakage, unintended visual cues, or changes in behavior caused simply by tethering an animal?

The movement also required hardware that could leave the microscope and enter the messy world of behavior. For many neuroscientists, the decisive emotional moment was not a voltage trace in cultured neurons but the first time a freely moving mammal could be controlled with pulses of light delivered through a fiber-optic interface. Historical accounts note the importance of such milestones: once light could be routed into the brain of an animal that was walking, exploring, freezing, or pressing a lever, optogenetics stopped being a clever physiology technique and became an instrument for circuit neuroscience. The fiber optic played an almost paradoxical role. It was an invasive implant, yet it enabled a form of control that was otherwise non-contact: the stimulus was light, not an electrode delivering current directly into the tissue.

It is easy to underestimate how much engineering hides in that apparently simple setup. Fibers must be chosen for core diameter and numerical aperture so that light spreads in a predictable way. Connectors must tolerate repeated handling. The light source must be stable and quickly modulatable. Commutators are needed so that an animal can turn without twisting a tether into a knot. Small design decisions become scientific decisions, because they affect how much tissue is illuminated and how reproducible stimulation is across animals. In a mature method, those decisions are not rediscovered in every lab; they are passed along as standards, improved incrementally, and treated with the same seriousness as any reagent.

Standards also emerged on the biological side. “Genetic targeting” became not a slogan but a discipline. It required careful matching of targeting strategy to experimental claim. Expressing an opsin in a brain region is not the same as expressing it in a particular neuronal population within that region, and expressing it in a population is not the same as stimulating only its cell bodies rather than its passing axons. As optogenetics spread, so did a vocabulary of caution: expression patterns must be verified; stimulation parameters must be reported; the possibility of indirect network effects must be respected. These norms are part of the movement's maturation. They are also how a method earns trust.

This is where the roles of different pioneers become visible without turning into hero worship. Edward Boyden's early work helped demonstrate that ChR2 could drive spikes on the brain's timescale with genetic specificity. Karl Deisseroth's group, and others alongside them, pushed the method into intact circuits and behavior while articulating the conceptual standard that optogenetics would be judged by: fast control plus genetic specificity, deployed to make causal statements about circuit function. Their contributions were not interchangeable. One could imagine a world where the initial demonstrations existed but the method remained confined to slices and cultures, or a world where in vivo demonstrations existed but were too idiosyncratic to reproduce widely. The movement required both the proof that the core actuator worked and the practical culture that made it portable.

Portability, however, did not mean uniformity. Once dozens of laboratories tried to implement optogenetics, limitations became obvious in ways that a single lab might not feel as sharply. ChR2 was powerful, but it was not perfect. Some experiments required faster closure kinetics for high-frequency patterns. Others required red-shifted sensitivity to penetrate tissue better. Still others needed inhibition rather than excitation, or the ability to modulate subthreshold activity without driving spikes. These demands fed back into protein engineering and comparative genomics. Reviews describe how the microbial opsin toolbox rapidly expanded beyond ChR2, yielding variants with different kinetics, spectra, and inhibitory or excitatory effects. This was not mere embellishment; it was a response to real experimental pain points.

That feedback loop is one of the signatures of a living technology. Neuroscience problems drove opsin engineering; improved opsins opened questions that had previously been impractical. A faster actuator can make it easier to mimic physiological firing patterns. A red-shifted actuator can make deep brain targets more accessible. A more light-sensitive actuator can reduce the required illumination and hence reduce heating or phototoxicity. Each improvement broadens the range of experiments that can be attempted, which brings in more users, which generates more edge cases and more demands for improvement.

The expansion of the toolkit also helped optogenetics avoid a common trap: becoming synonymous with a single molecule and therefore limited by that molecule's quirks. A movement cannot be pinned to one reagent. It must be modular. Once optogenetics became "opsins" rather than "ChR2," it began to look less like a one-off breakthrough and more like a general strategy: choose a light-responsive protein appropriate to the question, deliver it to the right neurons, deliver light with the right pattern, and interpret behavior with the discipline required for causal claims.

Modularity also changed how labs collaborated. In many subfields, a new method spreads through a slow, cautious diffusion: people visit the originating lab, learn the craft, and then rebuild it at home. Optogenetics spread more like open-source engineering. Constructs were shared. Detailed methods were published. Laboratories improved parts of the toolchain and sent those improvements back into the community. The result was a rapid iteration cycle in which many groups could contribute without each having to master every component. One lab might refine viral strategies; another might build better light-delivery hardware; another might engineer an opsin variant; another might develop behavioral paradigms that exploit the method's timing.

That style of dissemination also changed the sociology of training. A new graduate student could enter a lab with little molecular biology background and still learn to run an optogenetics experiment by following established protocols. That is a quiet revolution in expertise. It does not eliminate the need for skill, but it relocates the difficulty. Instead of spending years making the method work at all, students could spend their time using it to address biological questions, while still being held to high standards of validation and controls.

The movement also forced neuroscience to sharpen its thinking about causal interpretation. When a method is hard and rare, the community is often satisfied with demonstrations that it can be done. When a method becomes common, the emphasis shifts to what conclusions it legitimately supports. Optogenetics encouraged a more explicit separation between claims about sufficiency and claims about natural function. Driving a set of neurons and producing a behavior is a powerful sufficiency result, but it does not automatically mean those neurons play the same role naturally, under the brain's own patterns of activity. Conversely, silencing a set of neurons and preventing a behavior suggests necessity in that context, but it does not specify the computation those neurons perform. The method made these logical categories harder to ignore, because it offered tools that could probe them directly.

Even the basic act of "turning on" neurons turned out to have layers. Light delivered to a region can affect axon terminals from neurons whose cell bodies are elsewhere; stimulating terminals can have effects distinct from stimulating cell bodies. Light can recruit local inhibitory circuits and thereby suppress activity in neighboring populations indirectly. Animals can sometimes perceive light leaks or subtle heat, and can learn to associate those cues with stimulation. These are not embarrassments; they are the normal complications of bringing precise perturbation into a complex biological system. A movement earns credibility by making such complications discussable and by establishing best practices to detect them.

One can see the maturation in how papers began to report optogenetic experiments. Early excitement is often visible in the sheer fact that control is possible. Later work tends to sound calmer, because the method has become routine enough that authors must justify not the technique but the inference. How large is the illuminated volume? Which neurons express the opsin, and which do not? What is the latency from light onset to spiking, and does that latency drift over time? How does stimulation at one frequency compare with another? Does the behavioral effect persist when light is delivered in a control region or in animals lacking the opsin? These questions are the rhythm of a mature field.

Optogenetics also began to reshape what counted as a well-posed question. When an intervention is slow, it naturally pushes scientists toward questions about states: arousal, mood, general excitability. When an intervention can be timed precisely, it invites questions about events: the moment a memory is retrieved, the instant a fear response is initiated, the split-second when a cue biases a decision. That shift in temporal granularity is part of why the technique spread so rapidly across subfields. It offered a way to match perturbations to the brain's own temporal structure rather than imposing an external rhythm.

A movement, finally, is sustained not only by technical success but by a shared sense of possibility. Optogenetics created a feeling, in many labs, that previously untouchable questions could become tractable. Not easy, not automatic, but tractable: if you can control a selected neuronal population with the timing of a thought, you can begin to test models of circuits with an experimental rigor that was once reserved for simpler systems. That promise drew in engineers, molecular biologists, psychologists, and clinicians, each bringing different expectations and, crucially, different standards of evidence. The method survived that influx because it could be standardized enough to teach and flexible enough to adapt.

When a pulse of light can be delivered reliably to a defined set of neurons in a behaving animal, and when the genetic and optical parts can be swapped and improved without breaking the core logic, the technique stops being an isolated achievement. It becomes a shared instrument, one that many hands can tune, repair, and play.

Chapter 5

The Deisseroth Moment: When Psychiatry Met the Circuit Breaker

Psychiatry long described storms of feeling without being able to point to the wires that made them possible. Here the reader meets the moment when flashes of light began to turn anguish, fear, compulsion, and relief into testable circuit events—and when the harder question became not whether the brain causes behavior, but how to prove which cells matter, when, and in what direction.

5.1 The Symptom and the Wire

A psychiatrist learns early to take symptoms seriously, because they are the only things a patient can hand you directly. A person can say, with full authority, *I cannot sleep*, *My thoughts won't stop*, *I'm terrified for no reason*, *I feel nothing at all*. They can describe the way time slows down in depression, the way a panic attack ambushes the body, the way a craving in addiction feels like an argument you are losing in real time. But symptoms are also slippery. Two people can utter nearly the same list and yet carry entirely different stories inside the nervous system.

Clinicians notice this long before they have a name for it. One patient with "major depression" is weighted down by sleepiness and heaviness, eats more, moves slowly, and feels life as gray fog. Another with the same diagnosis wakes at 4 a.m. with dread, cannot eat, paces, and reports a harsh, self-accusing inner voice that sounds like certainty. The checkboxes overlap; the inner machinery might not. If you treat both with the same medicine, one improves, one doesn't, and the disappointment is not merely statistical. It lands as a question: what, exactly, is the "same illness" here?

Psychiatry, more than most of medicine, has been forced to make peace with categories built from lived experience rather than from direct measurement. A cardiologist can point to an occluded artery; an infectious-disease specialist can culture a pathogen. Psychiatry has brain scans and genetics, but no single signature that cleanly replaces the patient's account. The traditional diagnostic manuals are therefore careful and, in a way, humble: they do not claim to define a disease by its cause, only to group patterns of suffering that tend to travel together. That compromise has helped patients receive care and has helped research proceed, but it also sets a trap. When categories are defined by symptoms, symptoms can start to look like causes.

The mismatch shows up in everyday practice. A medication that eases panic in one person can leave another dulled and disengaged. A therapy that helps one patient with obsessive rituals can, in another, initially intensify distress as if the brain's alarm system has had its sensitivity turned up. The problem is not that the categories are useless; it is that they are coarse. A label like "anxiety" gathers many routes to the same destination: threat felt too easily, too strongly, or too persistently. The label describes the fire, not the wiring.

The "wiring" metaphor can mislead if it suggests something static, like a diagram etched in copper. Brains are not circuit boards.

They are living networks whose connections strengthen and weaken, whose chemical environments shift, whose rhythms drift with sleep and stress, and whose individual neurons can change their responsiveness depending on recent history. Still, the metaphor points to something real: mental life is built from communication between populations of cells. Feelings and thoughts are not stored in one place like files in a folder. They are patterns moving through routes, unfolding over time, with amplification, suppression, coordination, and competition.

Once that idea settles in, a symptom begins to look less like a single "thing" and more like a traffic pattern.

Consider fear. There is the kind you can justify: a car swerves into your lane, your heart leaps, your muscles tense, and the body's emergency chemistry surges in synchrony. But there is also fear without a fitting trigger, fear that lingers after the danger has passed, fear that attaches itself to safe places. Those experiences feel global because they recruit many systems at once: attention locks onto threat cues; the body prepares; memory searches for explanations; the social brain scans faces for signs of judgment or rescue. Yet global does not mean undifferentiated. The brain is achieving that total state by mobilizing particular pathways at particular moments. If those pathways are engaged too easily, disengage too slowly, or drown out competing signals, a person can be flooded by fear even while knowing, intellectually, that nothing is wrong.

The same logic applies to pleasure and motivation. "Reward" sounds unitary, but it contains components that can come apart: the enjoyment of a sensation, the learned pull of cues that predict it, the willingness to work for it, the feeling of satisfaction afterward. Addiction is often described as a failure of will, but from the circuit perspective it can look like a misallocation of priority: cues gain exaggerated access to motivational machinery; alternative goals lose influence;

habits become the default route. A craving, in this view, is not merely desire. It is a moment when one stream of neural messaging overwhelms others, like a microphone feeding back and dominating the room.

Even memory, which seems like a library function, is deeply about timing and coordination. A memory is not a photograph stored on a shelf. It is a pattern that can be reassembled when a cue arrives and a network locks into a particular configuration. If the wrong configuration is too easily triggered, intrusive memories appear. If the right configuration is hard to stabilize, concentration fails. When we speak casually about "short-term" versus "long-term," we are gesturing at processes that depend on how neurons synchronize their firing and how strongly one group can influence another at the right moment.

This way of thinking makes the old debates of "chemical imbalance" versus "psychological" feel oddly misplaced. Of course chemistry matters; neurons speak with chemicals and electricity. Of course experience matters; experience reshapes connections and expectations. The circuit view does not choose one side. It asks a different question: through which routes, with what timing, and with what gain does information flow?

Gain is a useful word borrowed from engineering: it means how strongly an input is amplified. Two radios can receive the same signal, but if the volume knob is higher on one, the sound dominates the room. Something like that happens in the brain. A threat cue may arrive at sensory systems in everyone; what differs is whether that cue is rapidly boosted into a body-wide alarm, or whether it is held in check by signals representing context, safety, and prediction. In depression, a negative interpretation might not merely occur; it might be given privileged amplification, while positive signals are weakened or quickly suppressed. Many psychotropic drugs, in one sense, can be seen as global gain-changers: they alter the chemical conditions under which circuits operate, making some pathways more influential and others less so. But global gain changes can be blunt. They can help, and they can also flatten, agitate, sedate, or produce side effects because they act across many circuits at once.

Timing is the other hidden variable. Brains run on rhythms. Neurons do not simply fire more or less; they often fire in coordinated bursts, in waves, in patterns aligned to ongoing oscillations that act like shared clocks. A signal that arrives at the right phase of a rhythm can be effective; the same signal arriving slightly off can be ignored or misrouted. This is one reason why psychiatric symptoms can feel state-dependent. Someone can function well for hours and then, with no obvious external change, tumble into obsession, panic, or despair. A small shift in internal timing can change which pathways dominate.

These ideas were in the air across neuroscience, but for a clinician trying to help a patient, they remained frustratingly abstract. The brain is not directly observable in the clinic. You can image it, but imaging often tells you about geography more than conversation: a region "lights up," a network seems connected or disconnected, a pattern correlates with a symptom score. Correlation can be informative, but it leaves a haunting ambiguity. If one circuit pattern accompanies anxiety, is it causing anxiety, responding to it, compensating for it, or merely traveling alongside it? A person's brain is always doing many things at once: regulating breathing, planning, remembering, anticipating. A scan that captures a broad blur of activity can miss the specific interactions that matter, like trying to infer the plot of a play from the temperature of the theater.

Karl Deisseroth entered this landscape with an unusual dual loyalty: to the patient's lived experience and to the mechanistic insistence of basic science. As a clinician, he faced the same puzzle that confronts psychiatry at large: diagnoses do not map neatly onto treatments, and symptoms that look similar can arise from different internal dynamics. As a scientist, he was trained to be dissatisfied with hand-waving. If a circuit is involved, which cells in that circuit are doing what, when? If a pathway is "overactive," what does overactive even mean in a code that uses spikes lasting milliseconds, in networks that coordinate with precision? The question was not just academic. If mental suffering is produced by specific patterns, then relief might come from correcting those patterns, not merely from bathing the brain in a general chemical tide.

A historical analogy helps. Before germs were accepted, doctors grouped fevers by how they looked and felt. They had categories, some useful, but they lacked the causal lever that could unite and separate cases. Once microbes could be identified, diseases that looked similar were split apart, and diseases that looked different were sometimes unified by a shared cause. Psychiatry's symptom categories are not "pre-scientific" in a pejorative sense; they are what careful observation yields when the causal machinery is hidden. The circuit view suggested a comparable shift: not from compassion to coldness, but from description to mechanism.

Mechanism, however, demands tools that the brain resists. If you want to know whether a specific neural pathway contributes to a feeling or an action, you need more than a snapshot. You need intervention: you must be able to perturb a defined component of the system and see whether a specific change follows. Neuroscience had ways to intervene—lesions, electrical stimulation, pharmacology—but each came with trade-offs. Lesions are irreversible and crude; they destroy more than they reveal. Electrical stimulation can be fast, but it excites anything nearby, including fibers of passage that simply traverse the area. Drugs can be targeted to receptors, but they diffuse, act slowly, and often alter many circuits simultaneously. These methods were powerful, and they built much of modern neuroscience, but they were poorly suited to the subtlety psychiatry needed: the ability to ask, in an intact brain, whether one particular kind of neuron in one particular projection, active for tens of milliseconds at one moment in a behavioral sequence, can tip the mind.

This is where the metaphor of a "wire" becomes precise enough to matter. In neuroscience, a "wire" is not a literal wire but an axon: the long extension of a neuron that carries spikes to other cells. Neurons of different types can intermix in the same region while sending their axons to different targets, like people living in the same neighborhood but commuting to entirely different jobs. A brain area can therefore participate in many functions, depending on which cell types and which projections are engaged. To say "the amygdala causes fear" is about as informative as saying "the airport causes travel." True at a coarse scale, but unhelpful if your problem is one delayed flight: you need to know which gate, which crew, which mechanical subsystem failed, and at what time.

A concrete case makes the stakes vivid: deep brain stimulation (DBS), a therapy in which electrodes deliver pulses to specific brain regions. DBS can be remarkably effective for movement disorders, and it has been explored for severe depression and obsessive-compulsive disorder. But DBS also dramatizes the circuit puzzle. Stimulating a site can relieve symptoms in some patients and do little in others, and even within one person the effects can depend on exact electrode placement. That sensitivity suggests that what is being influenced is not a whole "region" but a set of fibers and cell populations threaded through it. Millimeters can separate pathways with different roles. In other words, DBS hints—without fully explaining—that psychiatric symptoms may be circuit-specific in a way our usual labels do not capture.

The circuit reframing also changes how one thinks about comorbidity, the common reality that patients meet criteria for more than one disorder. From the symptom-list perspective, comorbidity looks like messy overlap. From the circuit perspective, it can be expected. One circuit motif—a particular pattern of miscommunication between, say, a control system and a threat system—could contribute to anxiety, depression, and addiction in different contexts. Conversely, two people with "depression" might have different motifs: one dominated by altered reward valuation, another by persistent threat prediction, another by impaired cognitive control. The brain does not respect the boundaries of our manuals; it respects connectivity and dynamics.

There is an ethical edge to all this. People often fear that mechanistic explanations will reduce them, turning anguish into circuitry and eroding responsibility or meaning. Yet the circuit view can do the opposite. It can replace moral language with understandable causes. A person with addiction who "should just stop" can be seen as someone whose motivational circuitry has been trained—by drugs, stress, and learning—into a state where cues have disproportionate influence. A person with obsessive-compulsive disorder who "knows it's irrational" can be seen as someone whose error-detection and habit circuits are locked in a loop that feels like unfinished business. These descriptions do not excuse harm, but they can reduce shame and focus attention on changeable mechanisms.

The great temptation, once circuits enter the story, is to talk as if one pathway "is" a symptom. That is rarely true. Circuits are reused. They are context-dependent. The same pathway can be helpful in one setting and harmful in another, like an emergency siren that saves lives during a fire but becomes torture if it blares all night. A circuit that promotes avoidance can protect you from danger; the same circuit, chronically over-engaged, can shrink a life. A circuit that energizes pursuit can build careers; the same circuit, hijacked by drugs or distorted valuation, can drive compulsion. What psychiatry needed was a way to move from plausible stories to disciplined tests:

to identify not just where something happens, but whether a specific cellular route is capable of producing a behavioral element, and whether that route is normally recruited when the behavior occurs.

Deisseroth's insistence on this level of specificity—on moving from symptom to wire—was not a rejection of the clinic. It was a refusal to accept that the clinic must remain permanently upstream of mechanism. The promise was not that every patient's suffering could be reduced to a single pathway, but that psychiatry could begin to trade in more exact explanations: not merely "too much fear," but "fear circuitry is being amplified at the wrong times, or safety signals are arriving too late, or control signals are too weak to compete." Those are hypotheses that can, in principle, be tested.

If this picture feels unsettling, it is partly because it suggests a kind of intimacy: that the most private parts of a person—fear, desire, hope, self-reproach—are enacted by physical messages in tissue. But it can also be comforting, because it implies that these experiences are not arbitrary, not signs of weakness, and not beyond understanding. They are constructed. And what is constructed can sometimes be reconstructed: not by denying the reality of suffering, but by learning which signals are being routed, amplified, or mistimed until the mind is cornered into a state it cannot easily leave.

5.2 No More Hand-Waving

A persuasive story about the brain has a particular feel to it. It often begins with a scan: this patch of cortex is quieter in depression; that deep structure is over-responsive in anxiety; this network looks disconnected in schizophrenia. The images can be beautiful and the statistics careful, and still a gap remains between *associated with* and *responsible for*. The gap is not pedantry. A circuit can light up because it is generating a symptom, or because it is trying to control one, or because it is collateral in a broader state change. Even when the correlation is rock-solid, the direction of influence can be wrong.

Optogenetics tightened that gap not by making the brain simpler, but by letting neuroscientists make claims that were previously too slippery to earn. The tool did not merely add a new way to poke the nervous system. It added a new standard: if you want to say a pathway helps produce a behavior, you should be able to intervene in that pathway with the right precision, prove that you did what you think you did, and show that alternative explanations do not quietly account for the result.

That last clause sounds obvious, but it is where many impressive-sounding brain claims have gone soft. Human beings are pattern-finders. When a behavior changes right after an intervention, our minds want to draw a straight arrow: the thing we touched caused the thing we saw. Brains, though, are full of hidden arrows. The intervention may have affected a neighboring population. It may have changed the animal's arousal, its motor readiness, its sensory processing, or its tolerance for discomfort. The behavior may have shifted because the animal was startled, or because the apparatus warmed the tissue, or because the experimenter, unconsciously, handled the animal differently. Strong neuroscience has to treat these possibilities not as cynical objections but as ordinary physics and ordinary biology.

The attraction of optogenetics was that it could, in principle, address the two biggest loopholes at once. First, it could reach inside mixed brain tissue and act on a defined population, rather than indiscriminately exciting everything near an electrode tip. Second, it could act on the brain at the timescale the brain actually uses. Thoughts and actions can unfold over seconds, but the signals that compose them are often coordinated at millisecond precision. If you perturb a circuit too slowly, you risk studying the brain's recovery from your manipulation rather than the brain's natural computations.

This is where Karl Deisseroth's clinician-scientist sensibility mattered. The clinical puzzle described earlier—the same diagnosis masking different internal traffic patterns—demands causal language, but it also punishes sloppiness. A psychiatrist cannot afford to confuse a sedative effect with a true reduction in fear, or a motor impairment with a reduction in compulsion. A patient might look "calmer" because they are slowed down, not because the internal alarm has quieted. In the lab, the same caution applies. If light makes an animal freeze, is the animal afraid, or is its movement circuitry disrupted? If the animal stops pressing a lever for a reward, is its craving reduced, or is its motivation intact but its motor output impaired?

Causation in a behaving animal is therefore not a single experiment but a style of reasoning. The basic vocabulary is borrowed from philosophy and engineering and made concrete in tissue.

One word is *sufficiency*. If activating a particular pathway at a particular moment can produce a behavioral element—a turn, a pause, an escape, a preference—then that pathway is sufficient to generate that element under those conditions. Sufficiency is powerful because it turns a vague association into a mechanism you can demonstrate.

The other word is *necessity*. If silencing a pathway prevents a behavior that would otherwise occur, then that pathway is necessary, at least in that context and at that moment. Necessity is often the stricter claim, because it forces you to confront the brain's redundancy: if the behavior still happens, the brain may be taking an alternate route. (The ability to silence cleanly is its own story, but even before inhibition enters, the discipline of sufficiency already demands rigor.)

Between sufficiency and necessity sits an often-overlooked point: a circuit can be sufficient to drive a behavior when artificially activated, yet not be the route the brain normally uses. This is not a failure of the method; it is a reminder that the brain has multiple ways to reach similar outputs. A car can be steered by the steering wheel, but you can also steer it by yanking the handbrake at speed; both change the car's trajectory, but only one is part of normal driving. Optogenetics made it easier to test sufficiency and necessity in a shared framework, but it also forced the field to become explicit about which claim was being made.

A historical anecdote shows how quickly this shift changed expectations. In 2007, Adamantidis and colleagues used light to stimulate hypocretin (also called orexin) neurons in freely moving mice and could trigger awakenings. The result was striking not only because it linked a defined population to a clear behavioral state, but because it linked them with timing. Wakefulness is not an abstract variable; it has an onset. If you can flip the relevant population at the right moment and the animal wakes, you have something closer to a lever than to a correlation. The experiment also made the field's obligation visible: if you claim to control wakefulness, you must show that you truly changed spiking in those neurons, that the effect is time-locked to the manipulation, and that the behavior is not explained by something trivial like light leaking to the eyes or a general stress response.

That obligation is the heart of "no more hand-waving". It is not about being harsh on new tools. It is about recognizing that a tool capable of dramatic effects is also capable of producing dramatic mistakes.

The logic of a strong optogenetic causal claim can be told as a story with checkpoints, each one designed to block a particular kind of self-deception.

You begin with a target that has a biological meaning: not "the left side of region X," but a population defined by its identity (for example, a genetically marked group) or by its projection (cells that send axons from one area to another). You deliver the opsin in a way that makes sense for the question. Then you verify expression, because biology is messy: viruses spread unevenly, promoters leak, and neurons vary.

Next comes a step that is easy to romanticize and easy to get wrong: you shine light and claim control. The mature field treats that moment as a hypothesis, not a triumph. Light does not come with a receipt saying "activated intended cells, no others." You must measure the effect of illumination on the targeted population. The gold-standard measurement is electrophysiology: recording spikes and showing that spikes appear (or disappear) with the timing and reliability you claim. Calcium imaging, which reports activity through fluorescent signals tied to calcium entry, can complement electrophysiology by letting you watch many cells at once and see whether the manipulation pushes the population into a new pattern rather than merely changing one neuron.

Only then do you align the intervention with a behavioral assay designed to separate the psychological construct you care about from simpler explanations. If you are studying anxiety, a behavior change could reflect fear, exploration, locomotion, or sensory sensitivity. If you are studying reward seeking, a change could reflect pleasure, effort, habit, or motor capacity. The assay has to be chosen with these ambiguities in mind, and often multiple assays are needed not to pad a paper but to triangulate meaning.

Finally, you earn the claim by doing controls that feel, at first, like killing the romance. You deliver light in animals that do not express the opsin. You implant fibers without illuminating. You match the pattern of stimulation in a way that rules out heating or photobiology.

You verify fiber placement because a millimeter can turn a "projection" experiment into a different circuit entirely. You quantify the light power at the tissue because "same laser setting" does not mean "same light at the target" across animals.

Without these steps, optogenetics can become a new kind of handwaving: the incantation of "we controlled the circuit" replacing the older incantation of "this region lit up."

Even without following the details, the moral is clear: the intervention is only as good as its validation, and the behavioral interpretation is only as good as its controls.

A concrete example from psychiatry-linked circuitry makes the point vivid. In 2011, Kay Tye and colleagues used projection-specific optogenetic manipulation within the amygdala to bidirectionally influence anxiety-related behavior. The amygdala is often treated as a monolithic "fear center" in popular accounts, but that shorthand collapses a dense set of interacting pathways. By stimulating a defined projection, the experiment moved past the broad claim that "the amygdala does anxiety" and toward a claim about a particular route. Just as importantly, the behavioral effect was rapid and reversible: it tracked the timing of illumination rather than drifting slowly the way many drug effects do. That rapid switching is exactly what can make causal reasoning cleaner, because it reduces the number of background variables that can change between "before" and "after."

But rapid reversibility also raises the bar. If a behavior flips quickly, then tiny confounds can also flip quickly. Light delivered through an implanted fiber can produce heat. Heating can change neuronal firing or alter nearby tissue in ways unrelated to opsins. Light can scatter to places you did not intend, including, in some setups, the retina or superficial sensory structures, which can startle an animal or change its arousal. Viral expression itself is not neutral: introducing a protein into neurons can stress cells or alter their properties; the surgery and the implant can cause inflammation or micro-damage; the mere presence of a tether can affect movement. These are not reasons to dismiss the method. They are reasons to do the dull, essential work of matching control conditions to the intervention.

This is where the culture of optogenetics began to resemble the culture of good pharmacology. A drug study cannot simply report that a pill changed behavior. It must report dose, route of administration, timing, and placebo controls. The optogenetic equivalent is to report light power, pulse pattern, opsin expression pattern, fiber placement, and "no opsin" controls. The field's maturation included formal guidance: detailed protocols in *Nature Protocols* (2010) and authoritative reviews in the *Annual Review* (2011) helped standardize what it meant to do the experiments responsibly. That kind of codification may sound bureaucratic, but it is one of the ways a powerful technique avoids becoming a generator of one-off miracles.

Timing deserves special emphasis because it is where optogenetics altered the conversation most sharply. If a circuit's role is to provide a brief signal—a gate that opens for a few tens of milliseconds, a teaching signal that updates learning, a synchronizing pulse that coordinates a network—then a slow intervention can miss the relevant event entirely. Earlier methods often forced scientists to argue indirectly. If a lesion reduced fear learning, did it remove the circuit that stores fear, or the circuit that carries the teaching signal, or the circuit that expresses the behavior? With time-precise control, you can intervene during learning, or during recall, or during expression, and ask which phase is sensitive. You can discover that a pathway is involved not "in fear" in general, but in the narrow window when a cue is linked to an outcome, or when a memory is reactivated.

Closed-loop experiments push this logic further. In a closed-loop design, the brain or the behavior triggers the light in real time. If a particular neural pattern appears, the system responds immediately, perhaps interrupting it or reinforcing it. That approach is closer to how symptoms feel: panic builds, peaks, and resolves; craving rises with cues; intrusive thoughts arrive uninvited. By intervening contingent on the state itself rather than on a fixed schedule, scientists can test whether a circuit pattern is merely accompanying the state or helping to sustain it. Closed-loop designs also reduce the temptation to over-interpret: if the effect depends on a specific detected state, it is harder to attribute the result to generic arousal or handling.

None of this makes neuroscience easy. It makes it honest. A strong optogenetic paper often reads, between the lines, like a series of arguments with oneself: *Could the light be doing something else? Could the virus be the true cause? Did we actually change spiking in the targeted cells, or did we merely assume it? Is the behavior a proxy for fear, or for movement?* The experiments become compelling not because they are theatrical, but because they anticipate these objections and answer them with measurements.

The payoff is not a single "fear circuit" or a single "depression circuit," but a new kind of explanatory precision. A symptom becomes something you can parse into components that unfold over time, each potentially linked to a definable route and a definable code. When the claim is earned, it has a particular kind of solidity: not that the brain has been reduced to a switch, but that a piece of the wiring has been touched with enough care that the resulting change in behavior can be taken seriously as evidence about how that wiring participates in the mind.

5.3 The Power to Say No

Turning something on is an invitation to overconfidence.

If you activate a pathway and an animal freezes, it is hard not to feel you have found "the fear circuit." The story seems neat: press the button, get the behavior. Yet brains are generous to experimenters in a slightly mischievous way. There are many routes to immobility: fear, pain, surprise, motor interference, even a shift in attention that makes movement less likely. Artificial activation can also recruit downstream systems in ways the brain rarely uses on its own. A circuit may be *capable* of producing a behavior without being the circuit the brain normally relies on to do it.

That is why optogenetics became fully convincing only when it learned to say no.

In everyday life, "no" is a boundary. In circuit science, "no" is a test. When you can turn off a defined population, briefly and reversibly, you can ask a stricter question than sufficiency: is this element required at the moment the behavior is naturally generated? If the behavior collapses when the element is silenced, the causal claim tightens. If the behavior persists, the story must change. Either way, the brain has forced you to be honest.

The practical problem is that neurons are not eager to be switched off cleanly. Traditional tools could silence, but they tended to be blunt. Lesions destroy tissue permanently and provoke compensations that unfold over days or weeks; the brain reroutes, adapts, and sometimes learns around the missing piece. Drugs can reduce firing, but they diffuse and linger, affecting many targets for minutes to hours. Even techniques that "inhibit" locally can inadvertently excite other fibers or change blood flow. For the kind of circuit questions psychiatry demands—questions about timing, about specific projections, about fleeting internal states—these older methods were often like trying to understand a conversation by cutting the power to the whole building.

The optogenetic contribution was not merely that light could trigger spikes. It was that light could be used to suppress spiking with comparable precision. In 2007, Zhang and colleagues demonstrated optogenetic inhibition using microbial halorhodopsin alongside excitation, and the conceptual landscape shifted. Now the same general approach—genetically specified expression plus light delivery—could be used for both directions of control. A circuit diagram could be interrogated like an electrical engineer interrogates a system: not only "can we drive current through this wire?" but also "what stops working if we open the switch?"

Halorhodopsin, in its optogenetic form, is a light-driven pump that moves chloride ions into the neuron. Arch (short for archaerhodopsin) is a light-driven proton pump that moves protons out. Both tend to push a neuron away from firing, making it less likely to produce spikes while the light is on. The exact biophysics can be intricate, but the experiential point is simple: the light becomes a reversible muzzle. When the illumination stops, the neuron can return to its prior behavior, often within seconds. That reversibility is crucial, because it allows the same animal to serve as its own control: light on versus light off, within one session, without waiting for the brain to rewire itself around a permanent injury.

This kind of silencing changed the tone of circuit claims. Early optogenetic activation experiments could be dazzling, like stage magic: a pulse of light and a behavior flips. Inhibition demanded a more sober posture. If you silence a pathway and nothing obvious changes, you must resist the temptation to call the pathway irrelevant. Perhaps the behavior is supported by redundant routes; perhaps your timing was off; perhaps your silencing was incomplete; perhaps the animal's strategy changed. A negative result can be informative, but only if you can prove you truly suppressed spiking in the intended neurons and did not simply miss the cells or underpower the light.

That requirement brought the field back to the discipline described earlier: verification, control conditions, careful behavioral design, and an explicit statement of limits. Silencing can make the causal logic tougher, but it does not magically remove confounds. Light delivery and implants can themselves alter network state and behavior. Viruses can perturb cells. Tissue around a fiber can change over time. A "no" is only meaningful if you know that the circuit element was actually denied its usual contribution.

The deeper conceptual shift, though, was psychological. Turning circuits on tempts you to treat the brain as a set of buttons. Turning circuits off forces you to treat it as a system with dependencies.

A helpful analogy comes from aviation. If you push an aircraft's controls in a way a pilot never would, you can make it do dramatic things, including dramatic failures. That does not tell you how the plane normally flies. But if you disable a specific instrument midflight and the pilot immediately loses a capacity—cannot hold altitude, cannot navigate—then you have learned something about what that component is normally doing. Circuit inhibition asks the brain a similar question: during natural behavior, which elements are acting as instruments without which the system cannot maintain its function?

This is exactly the sort of question that resonates with psychiatry, where symptoms often feel like a failure of control. A person with addiction may describe watching themselves reach for a substance as if a veto has been removed. A person with obsessive-compulsive disorder may describe a compulsion as something they do not endorse but cannot stop. A person with panic may feel a bodily alarm that reason cannot override. These are not metaphors about "wiring" used for comfort; they are descriptions of a lived experience of agency being altered. If a circuit is truly providing a veto, a brake, or a gate, then turning it off at the right time should have consequences.

The addiction literature provided an especially sharp demonstration because it offered something rare in psychiatry: a clear behavioral readout that can be measured repeatedly, and a condition—compulsive drug seeking—that seems to involve a loss of control rather than a simple increase in pleasure. In 2013, Chen and colleagues reported that in a model of compulsive cocaine seeking, animals showing the most compulsive behavior also showed prolonged hypoactivity in the prelimbic region of the prefrontal cortex. "Hypoactivity" can sound vague until you translate it into a circuit hypothesis: a region thought to help exert control over action selection and evaluation is not providing its usual influence.

Optogenetics let the authors test that hypothesis in both directions. Stimulating prelimbic cortex could rescue the behavior, preventing compulsive cocaine seeking in the model. Inhibiting prelimbic cortex could push behavior toward compulsion; silencing was sufficient to promote the very pattern the stimulation could reduce. The bidirectionality is what makes the result more than a one-way trick. When pushing a circuit up shifts behavior in one direction, and pushing it down shifts behavior in the opposite direction, it begins to look like a control knob the brain normally uses.

Even here, silencing adds a crucial nuance. If activation alone reduced compulsion, a skeptic could argue that the stimulation produced a general arousal, or disrupted habitual responding, or made the animal less able to perform the seeking behavior. Inhibition, by contrast, tests whether removing that prefrontal contribution makes the animal *more* compulsive, which is harder to explain as a trivial motor impairment. The same reasoning holds across domains: silencing often helps separate "we interfered with behavior" from "we removed a specific regulating influence."

The emotional stakes of this are easy to miss if one reads these studies as mere mouse behavior. In the clinic, "self-control" is a morally charged phrase. Patients are blamed for lacking it, praised for having it, judged by it. A circuit account does not absolve responsibility, but it does relocate the discussion. If a particular prefrontal influence is weakened—by drug exposure, stress, genetics, or development—then the felt experience of "I can't stop" is not simply a confession of bad character. It is a report that the brain's braking system is not exerting normal leverage. That reframing can support compassion without surrendering to fatalism, because circuits can change. They can be strengthened, retrained, or supported by interventions that are more precise than blanket sedation.

Silencing also clarified another common confusion: the brain region versus the brain pathway. A region is a crowded marketplace. It contains multiple neuronal populations, some excitatory, some inhibitory, some projecting to one target, some to another. If you "turn off" a region with a drug, you may be silencing neurons that normally oppose each other. The net behavioral change can be small or paradoxical. Optogenetic inhibition, especially when targeted to a defined projection, can address this. Projection-specific silencing can deny influence from area A to area B while leaving other outputs of A intact. This helps transform statements like "the amygdala drives anxiety" into testable claims about particular routes.

Anxiety circuitry provided an arresting public-facing example of reversible control: stimulating a specific amygdala projection could produce acute anxiolytic effects that reversed when illumination stopped (Tye and colleagues, 2011). Although that particular demonstration is often told as a triumph of activation, its deeper implication is about structure: what looks like one emotional state can be tipped by adjusting the traffic through one defined connection. Once the field had reliable inhibition tools, the same logic could be asked in the opposite direction: does removing that connection's influence at a critical time restore anxiety, prevent relief, or leave behavior unchanged because other routes compensate? Those are not merely technical refinements. They are the questions that determine whether a circuit is a lever the brain normally pulls, or merely a handle an experimenter can grab.

"Critical time" is not rhetorical; it is literal. Some circuits contribute only during particular phases. A pathway may be required for learning but not for expression, or for expression only when a cue appears, or for maintaining a state once it has begun. Inhibition makes these temporal roles testable. You can silence for a second, during cue presentation, and leave the rest of the task untouched. If behavior changes only when the inhibition overlaps a specific event, then the circuit's role becomes more precisely located in time. This is particularly relevant for psychiatric symptoms that arrive in bursts—panic spikes, intrusive memories, craving surges—where the subjective experience is that something hijacks the mind for moments and then releases it.

There is, however, a trap that inhibition introduces, and it is worth naming because it keeps the field from replacing one kind of overclaim with another. When you silence a circuit element and behavior changes, it is tempting to say you have found "the necessary circuit." But the brain is not obligated to use one route forever. Necessity is conditional. It depends on the animal's strategy, the environment, developmental history, and the broader network state. Silencing may produce a behavioral change because you removed the intended influence, or because you created an imbalance that propagates through the network in a way that is not identical to the natural state. Optogenetic inhibition is fast and reversible, but it can still be an unnatural event. It denies the brain the opportunity to transition gradually, and it can lead to rebound effects when the light turns off.

That is why claims of "saying no" at the circuit level have to be paired with the same confound-awareness that activation demanded.

The light might heat. The implant might irritate tissue. The opsin protein might subtly alter the cell. The act of silencing might shift local chemistry in ways that influence neighboring circuits. Good experiments match light delivery and implants in controls, verify suppression in the targeted neurons, and report parameters transparently. The strength of inhibition is not a philosophical guarantee; it is a measured effect.

Still, when done carefully, silencing offers something close to what clinicians crave: a way to link a symptom-relevant behavior to a definable circuit element with the logic of removal. It is one thing to show that activating a route can evoke freezing, lever pressing, or avoidance. It is another to show that without that route's contribution—without that particular message arriving at that particular time—the behavior loses its normal form.

For psychiatry, the most hopeful aspect of this is not the fantasy of turning symptoms off with light. It is the prospect of making the right distinctions. Two patients may look the same on a diagnostic checklist, yet rely on different circuit motifs. One may have a threat system with its gain turned up; another may have a control system that fails to apply brakes at the decisive moment. Activation can show you what a circuit *can* do. Inhibition can show you what the circuit is *doing* when the organism is itself, trying to cope, decide, avoid, pursue, remember.

A clinician listening closely to a patient often hears a similar logic in ordinary language. "I know it's not rational, but I can't stop." "I want to quit, but when the craving hits, it's like something else takes over." "I can talk myself down most days, but sometimes my body refuses." These are descriptions of vetoes that fail, gates that stick, amplifiers that blare. Optogenetic silencing turned those descriptions into experimentally tractable questions, and it did so in the most stringent way science knows: by removing a candidate component and seeing whether the system can still do its job.

5.4 Open Circuits, Shared Standards

A tool that can make an animal freeze on command, or reduce anxiety-like behavior with a pulse of light, carries a predictable risk: it invites theater. The results are vivid, the videos easy to share, the metaphors irresistible. In neuroscience, vividness has always been a double-edged gift. It draws attention, funding, and talent; it also tempts a field to outrun its own checks and balances. Optogenetics could have become exactly that kind of phenomenon: a dazzling trick that generated more headlines than reliable knowledge.

It did not become that, at least not in the main. One reason is technical: microbial opsins proved robust, flexible, and engineerable. Another reason is cultural: the technique spread through a kind of deliberate openness that functioned as quality control. Protocol papers, shared plasmids, standard parts, and blunt conversations about artifacts did not merely accelerate adoption. They created shared standards for what counted as a believable causal claim.

This is an unusual ending for a story about a biological invention, because it treats a community's habits as part of the invention. Yet the social infrastructure mattered as much as the optical hardware. Without it, optogenetics might have been trapped inside a handful of expert laboratories, or worse, diffused so rapidly and idiosyncratically that every laboratory would have its own private version of "optogenetic stimulation," making results hard to compare and easy to misinterpret.

A timestamp captures how quickly the technique moved from specialist tool to communal platform. By around 2010, optogenetics had become a mainstream, cross-field method; *Nature Methods* named it Method of the Year in 2010. That kind of recognition reflects more than popularity. It signals that a technique has started to set expectations across subfields, from sensory physiology to psychiatry-relevant behavior. The moment a method becomes a lingua franca, its weaknesses become everyone's problem. What had been a clever solution in one corner of neuroscience now had to survive contact with thousands of experiments, performed under different assumptions and levels of expertise.

The first task in that situation is surprisingly mundane: making the method portable. If a graduate student in a new laboratory cannot reproduce the basic effect, the technique cannot serve as a platform. Portability required standardizable protocols and shared reagents. Detailed protocols in *Nature Protocols* (2010) played a crucial role here. A protocol paper is not glamorous. It is the scientific equivalent of a well-written manual: how to deliver the opsin, how to implant the fiber, how to calibrate light, how to avoid common failure modes, how to confirm expression. Protocols turn tacit expertise into something that can be taught and audited.

At the same time, consolidating reviews in the *Annual Review* (2011) and Karl Deisseroth's 2011 overview in *Nature Methods* did something equally important: they stabilized the vocabulary. They helped a rapidly growing field agree on what questions optogenetics is good for, what caveats travel with the method, and what experimental logic should be considered persuasive. This is not mere pedagogy. In a young technique, language is part of the technology. If "activation" means one thing in one laboratory and another elsewhere, the field cannot accumulate knowledge; it can only accumulate anecdotes.

The second task is more delicate: making the method shareable without making it chaotic. Here, plasmids became as important as lasers. A plasmid is a circular piece of DNA used in molecular biology as a vehicle for genes; it is a way to package an opsin with the genetic switches that control where and when it is expressed. If you can send a plasmid to another laboratory, you are not sending a description of the tool; you are sending the tool itself, with its sequence, its promoter, its tags. That matters because, in biology, small changes in sequence or expression strategy can yield big changes in results. A technique built on genetic targeting can fragment into incompatible variants if everyone rebuilds it from scratch.

Material-sharing infrastructure made this less likely. Repositories such as Addgene, which distributes plasmids, allowed optogenetics plasmid collections to be shared at scale. This is one of those behind—the-scenes systems that looks boring until you imagine science without it. Without repositories, sharing happens through personal networks and informal favors. That kind of sharing is slow, uneven, and subject to gatekeeping. With repositories, a laboratory in another country can start from the same baseline construct used in a famous study. Replication becomes feasible not only for well-connected groups but for anyone with the skill and the patience.

Openness also lowered a subtle barrier: the fear of being "scooped" by one's own generosity. When a field's leaders share reagents and methods early, they normalize a kind of confidence. They implicitly say: the value is not in hoarding a plasmid, but in asking good questions and answering them rigorously. Karl Deisseroth and Ed Boyden both played visible roles in accelerating uptake in this way, not only through papers but through the circulation of tools and the willingness to engage with other laboratories' troubleshooting. The result was not that credit disputes vanished—science remains human—but that the method itself did not become a private dialect spoken only by its inventors.

There is an additional, less obvious benefit to openness: it makes hype easier to resist. A dramatic behavioral effect can be compelling, but it can also be fragile. If only one laboratory can produce it, the field cannot easily decide whether the effect is a deep truth or an artifact. Shared tools make it easier to check claims quickly. If a paper reports that stimulating a pathway reduces anxiety-like behavior, other groups can test the same opsin, similar light parameters, and comparable targeting strategies. If the effect replicates across conditions, confidence grows; if it does not, the conversation becomes more nuanced. In this sense, openness is not a moral stance but a way to keep scientific claims tethered to reality.

The need for tethering was real, because optogenetics comes with confounds that are not always intuitive. Light is not only a trigger for opsins. It can warm tissue. It can cause photobiological effects. It can leak and alter an animal's perception. Viral delivery can perturb cells and immune responses. The implant can damage tissue or change local glial activity. These issues do not make optogenetics unreliable by default, but they mean that a visually impressive result is not self-validating.

As these interpretive pitfalls became more visible, stronger norms emerged for controls and parameter reporting. A field-wide method needs field-wide discipline: matching controls for light delivery, viral expression, and implant effects, not merely showing that behavior changes when illumination is applied. That discipline does not arise spontaneously. It is cultivated through protocols, reviews, reviewer expectations, and, importantly, the ability of other laboratories to reproduce or fail to reproduce a result using comparable reagents.

A concrete anecdote, familiar to many experimentalists, shows how this culture works in practice. Imagine a laboratory newly adopting optogenetics after seeing a striking paper. They order a plasmid from Addgene, follow a *Nature Protocols* guide, build or buy a light source, and perform the surgery. Their first experiment yields an effect, but it is weaker than expected, or it appears only in some animals, or it comes with a puzzling increase in general arousal. At that moment, the openness of the field becomes a kind of diagnostic system. Because other groups are using similar parts, the laboratory can compare notes: light power at the fiber tip, expression pattern, timing, behavioral assay, animal handling. The troubleshooting is not simply technical; it is epistemic. It helps distinguish a true biological boundary condition (the effect depends on state, sex, strain, or context) from a methodological error (misplaced fiber, insufficient expression, excessive heating).

That sort of troubleshooting can feel like the opposite of discovery, but it is where a method becomes reliable enough to support claims about the brain. A field that shares tools also shares failure modes. Over time, these failure modes become common knowledge. They show up in talks not as embarrassing footnotes but as expected wisdom: yes, heating can confound results; yes, light can cause artifacts; yes, viruses can do more than deliver a gene. The collective memory of pitfalls becomes part of the technique's maturity.

There is a deeper point here about psychiatry. The disorders that motivate much of this work are not tidy. Depression is heterogeneous; anxiety has multiple faces; addiction is intertwined with learning, stress, and social context. A technique that produces clean, time-locked behavioral changes can seduce researchers into telling overly clean stories. Shared standards push back. They demand that an effect be interpreted within a broader set of measurements: electrophysiological verification, histological confirmation, and behavioral specificity. They encourage the field to state not just what a circuit manipulation can do, but what it does *not* do, what it fails to change, and what alternative explanations have been ruled out.

As optogenetics spread, it began to resemble other successful scientific platforms: PCR in genetics, CRISPR in genome editing, MRI in human imaging. Platforms create power by standardizing. They also create responsibility, because standardized power can be scaled.

When any laboratory can, in principle, manipulate a circuit and observe a behavior, the community must be able to trust that "manipulate" and "observe" mean roughly the same thing across contexts. That is why parameter reporting matters. It is why shared controls matter. It is why reviewers and editors gradually learned to ask for details that would have sounded fussy in the early days: light power density, stimulation pattern, expression spread, fiber placement, verification of effect on spiking, and matched controls for light and implants.

Openness, in this sense, is a form of calibration. A laboratory does not only calibrate its lasers; it calibrates its claims against a community that can check them.

The ethical implications of this are not theoretical. The more a tool is described as a "switch," the more it invites fantasies of direct control over the mind: turning off sadness, turning down craving, erasing fear. Those fantasies can be seductive, especially to people who suffer and to societies that want quick fixes. Optogenetics, used responsibly, pushes against that kind of simplification. Its strongest experiments do not show that a single pathway "is" an emotion. They show that specific patterns of signaling contribute, at specific times, in ways that can be tested and bounded.

That boundary-setting is not only good science; it is good citizenship. If the public hears only that scientists can "control behavior with light," the story becomes one of manipulation and loss of agency. If the public also hears that the method demands rigorous controls, that effects depend on precise targeting, that artifacts exist and are actively policed, then the story becomes more accurate: brains can be studied with causal tools, but causal tools are not magical, and their meaning is earned through shared standards.

One can see the mature form of this attitude in later critiques and methodological discussions that treat optogenetics not as a finished miracle but as a technique with known vulnerabilities and known remedies. The existence of such critiques is a sign of health. A field that cannot criticize its own flagship method is a field at risk of turning technique into ideology.

What emerged, then, was a kind of circuit science that could travel. A laboratory studying memory could borrow a construct and a protocol from a laboratory studying fear. A group interested in addiction could adopt a stimulation pattern validated in an arousal study, while adapting controls to its own behavioral assays. This portability did not flatten the science; it made it cumulative. Each experiment could become not an isolated spectacle but a data point in a growing, shared map of how timing, gain, and communication in neural circuits shape the mind.

When patients eventually benefit from circuit-level insight—whether through better drugs, better stimulation therapies, or more precise psychological interventions—they will not be benefiting from a single invention. They will be benefiting from an ecosystem: shared reagents, transparent protocols, and a community that learned to treat extraordinary claims about the brain as invitations to replicate, to quantify, and to argue in public until the claims either hardened into knowledge or softened into humility.

Chapter 6

Getting the Switch Into the Right Cells: The Art of Targeting

A flash of blue light means very little if it lands on the wrong cells. Beneath every elegant optogenetic experiment lies an invisible logistics operation—engineered viruses, genetic gatekeepers, atlas-guided surgery, and skeptical verification—that turns a vague brain region into a sharply defined target.

6.1 Viruses as Tiny Couriers, Not Tiny Villains

The word *virus* lands with a thud. It arrives carrying the emotional weight of pandemics, childhood fevers, and news headlines. In optogenetics, though, a virus is usually closer to a FedEx envelope than a burglar. It is a delivery vehicle that biologists have stripped down, re-engineered, and put on a tight leash so it can do one job well: smuggle a small piece of genetic text into a cell so that the cell can manufacture a light-sensitive protein.

The workhorse for this job is adeno-associated virus, often shortened to AAV. In the wild, AAV is a small virus that cannot even replicate efficiently on its own; it typically needs help from a second "helper" virus to multiply. In the lab, the version used for neuroscience is more domesticated still. The genes that would normally allow it to reproduce and spread are removed and replaced with the experimenter's cargo. The result is a shell that can enter cells and deliver DNA, but is engineered to be replication-incompetent under normal laboratory conditions. When people say "we injected an AAV," they are usually describing a carefully controlled reagent: made in cultured cells, purified, quantified, and handled under institutional biosafety rules that treat it very differently from a pathogen circulating in nature.

That domestication is not a hand-wavy reassurance; it is the entire reason AAV became so popular. AAV tends to be relatively gentle on neurons, and it can drive gene expression for weeks to months, which is exactly what is needed to do behavioral experiments that involve training, stress, addiction paradigms, or chronic disease models. In a practical sense, AAV also fits the rhythms of a lab: you can plan an injection, wait for expression to build, implant an optical fiber, and then run experiments with a stable preparation.

Yet it helps to hold two thoughts at the same time. First: these vectors are engineered to reduce risk and increase usefulness. Second: delivery is biology, not plumbing. The virus is not a neutral pipe. It has preferences about which cells it enters, how far it travels from the injection site, how strongly it expresses the gene it carries, and how the host tissue reacts. Those preferences become part of the experiment's logic, whether the experimenter wants them to or not.

AAV's appeal comes with a tight constraint: the payload is small. Think of the capsid (the protein shell) as a suitcase with a strict airline weight limit. An opsin gene by itself can fit comfortably, as can a fluorescent tag that lets you see where the opsin ended up. But the moment you start adding regulatory elements that make expression stronger or more specific, you can hit the packaging ceiling. That tradeoff shapes the entire field. Many of the elegant genetic tricks people associate with optogenetics are, in part, clever ways of squeezing instructions into limited space.

The limitations are not only about *what* you can pack, but also *how* it will behave once it arrives. AAV vectors come in different "serotypes," which are best understood as different outer coats. The coat determines how the virus interacts with the surfaces of cells and how readily it is taken up. In neuroscience conversations, you may hear a serotype name said with the casual confidence of a chef talking about olive oil: "We used AAV9," or "We switched to AAV5 for better spread." This shorthand can sound like superstition until you appreciate what it represents: a set of accumulated observations about tropism, the tendency of a vector to infect some cell types more readily than others, and to move through tissue in particular ways.

Tropism is not destiny, but it is a bias. One AAV serotype may preferentially transduce neurons over glia in a certain context; another may spread more widely from the injection site; another may work better in one species than another. Even within "neurons," the bias can be subtle. The physical layout of cells, the density of extracellular matrix, and the patterns of receptors on cell membranes all influence uptake. AAV is a molecular object moving through a crowded forest, and each forest is a little different.

A vivid example of how far this engineering can go arrived in 2016, when Tervo and colleagues reported a designer AAV variant optimized for retrograde access to projection neurons (Tervo et al., 2016). "Retrograde" here means traveling back along the wiring: infecting axon terminals in one location and delivering the genetic payload to the cell bodies that project there. That capability is powerful because it begins to define a population of neurons not only by where they live, but by where they send information. The key point for targeting is not the specific capsid variant; it is the principle that the viral shell itself can be tuned like a tool, and that tuning changes what "the targeted neurons" even means.

If AAV is the courier, the experimenter still has to choose the route and the package size, and those decisions show up later in the data.

Start with dose. In AAV work, "dose" is not a pill; it is a concentration of viral particles, often discussed as a titer, and a volume delivered. More particles generally increase the chance that cells near the injection will be transduced and increase the number of DNA copies inside each cell. That often yields brighter fluorescence and stronger opsin expression. It can also produce more off-target expression at the edges of the intended zone, and it can increase immune responses or tissue irritation, depending on the context. Too low a dose and the experiment becomes a dim lightbulb: the opsin is present, but not enough to reliably drive or silence spikes. Too high a dose and the experiment can become something else entirely: a broader manipulation than intended, or a perturbation that changes baseline physiology.

Injection volume is dose's inseparable partner. A larger volume can help fill a structure and increase the number of transduced cells, which is tempting when a nucleus is big or when a behavior seems to require recruiting enough neurons to have an effect. But fluid injected into brain tissue is not a gentle mist; it is a mechanical event. A large bolus can damage tissue, physically displace cells, and create backflow up the needle track. Backflow is the unglamorous enemy of targeting: the vector solution takes the path of least resistance and travels along the track you created, smearing expression toward the surface or into adjacent areas you never meant to touch. Many labs therefore inject slowly, sometimes leaving the needle in place after injection to let pressure equalize. These choices are practical craft, yet they shape the biology of the final preparation.

Even with identical dose and volume, spread varies by tissue properties. Some areas of the brain are tightly packed; others are more permissive. White matter tracts, rich in myelinated axons, can behave like highways for fluid movement or barriers, depending on geometry and injection angle. The extracellular matrix can trap or channel viral particles. If two neighboring structures have different cellular densities, the same injection can produce a sharp boundary on one side and a fuzzy gradient on the other. And because individual brains vary, the same coordinates can land in tissue that is slightly more vascular, slightly more scarred from an earlier surgery, or slightly different in developmental history, changing how the virus disperses.

This is where the intuition of "a virus delivers a gene to *the* region" starts to break down. The reality is usually a cloud of expression: dense near the injection site, thinning outward. Within that cloud, not every cell takes up the payload. Among those that do, not every cell expresses it at the same level. Some cells may express the opsin so strongly that a small amount of light can drive robust currents; others may be weakly transduced and barely respond. If you picture optogenetics as flipping a crisp, digital switch, AAV delivery reminds you that biology prefers analog gradients.

These gradients have consequences for interpretation. Suppose a researcher injects an opsin into a structure thought to control fear learning. The behavioral effect of light might be driven by the densely infected core, by a sparse halo, or by an unintended spill into a neighboring pathway. If the injection happens to nick fibers of passage (axons traveling through the area), the virus might infect axon terminals and be transported to distant cell bodies, depending on the vector's properties. Suddenly the manipulation is not purely local even if the fiber optic is.

In the everyday language of papers and presentations, these complexities can get compressed into a single phrase: "virus expression was confined to the target." That phrase can be honest in a broad sense and still hide an uncomfortable truth: *confined* often means "mostly concentrated," not "exclusive." The brain is not divided by walls. It is divided by gradients and interleaving cell types, and the vector's biology participates in those gradients.

The choice of serotype adds another layer. AAV serotypes differ in how readily they infect neurons versus glia, how strongly they express in certain contexts, and how far they spread. Those differences are not just technical trivia; they are the experimental equivalent of choosing a paintbrush. A brush that lays down thick paint quickly is useful if you need strong opsin currents. A brush that makes a wide, light wash is useful if you want broad coverage. Neither is "better" in the abstract. Each one changes the causal claim you can make.

Imagine two labs studying the same behavioral phenomenon, such as compulsive seeking after drug exposure. Lab A uses a serotype that produces tight, dense expression near the injection site, and they place the fiber directly above that region. Lab B uses a serotype known for broader spread and injects a larger volume to "cover the area," then places a fiber at a slightly different angle. Both might say they manipulated the same structure. Both might even see a change in behavior with light. Yet the set of neurons actually expressing opsin could differ dramatically, and so could the interpretation. One lab may be primarily affecting local cell bodies; the other may be influencing a mixture of local neurons and fibers that pass through or terminate nearby. When results disagree, the explanation is not always deep neurobiology; sometimes it is logistics.

AAV's small payload limit also shapes what gets delivered along with the opsin. Strong expression often requires regulatory elements that enhance transcription or stabilize mRNA. Add too many, and the construct no longer fits. So researchers make compromises: a shorter promoter, a compact opsin variant, a minimal tag. Each compromise shifts the balance among brightness, specificity, and practicality.

The word "promoter" may sound forbidding, but the idea is simple: it is a stretch of DNA that acts like a start signal for gene expression, helping determine where and how strongly a gene is turned on. Promoters can be chosen to tilt expression toward neurons in general or toward certain classes of neurons. But promoters are not magic labels; they are snippets of regulatory language borrowed from biology, and outside their natural genomic context they can behave unpredictably. Put a promoter in an AAV and it may be less specific than hoped, because it lacks distant regulatory elements that normally refine its expression. Or it may be weaker than expected in a particular region or species. The same promoter that seems beautifully selective in one dataset can look messy in another.

All of this can make AAV sound like a finicky tool, and in a sense it is. But finickiness is the price of working at the scale of individual cells inside a living brain. The deeper point is that AAV's "personality" is not a nuisance to be ignored; it is part of the experimental design.

Consider the most common practical question a lab faces: *Do we want local precision or broad coverage?* If you are asking whether a tiny nucleus is sufficient to trigger a panic-like response, you may want tight expression and a small injection to avoid touching neighboring structures. If you are asking whether a distributed population contributes to a slow mood-like state, you might accept broader spread to capture enough of the relevant neurons. These are not merely technical decisions; they are commitments about what kind of causal statement you are trying to make.

There is also a quieter, human story behind AAV's dominance: trust built through repetition. Early gene therapy history included sobering moments, including severe immune reactions in some viral-vector trials. Those episodes were not about AAV in neuroscience, but they shaped the culture: skepticism, biosafety review, and a recognition that vectors are powerful biological agents. AAV gained credibility in part because it proved, across many studies and contexts, to be comparatively well-tolerated while still effective. In clinical medicine, AAV's rise has been tied to the promise of delivering genes to treat inherited disorders, where long-lasting expression is a feature, not a side effect. In basic neuroscience, that same stability allowed experiments that unfold over weeks: training animals, measuring long-term changes, and returning to the same circuit again and again.

Yet "stable" does not mean "invisible." The immune system can notice viral capsids; cells can respond to stress; expression levels can drift; and different animals can react differently. The brain's response is often subtle, but subtle shifts can matter in experiments that hinge on behavior. This is one reason why optogenetic work treats controls with such seriousness: a control virus that expresses only a fluorescent protein, or light delivered in a way that should not activate the opsin, helps separate the effect of the optogenetic switch from the effects of surgery, viral exposure, and illumination.

The romance of optogenetics often focuses on the light: the fiber optic, the laser, the crisp timing. The unromantic truth is that much of the craft lives upstream, in the choice of courier and the details of delivery. AAV is powerful precisely because it can be tuned. That tuning is not an afterthought. It is the difference between "we changed activity in a particular place" and "we changed activity in a particular set of cells," and between a claim that will hold up under scrutiny and one that dissolves when someone repeats the experiment with a slightly different serotype, dose, or injection volume.

AAV, then, is not a tiny villain and not a neutral messenger. It is a domesticated biological machine: reliable enough to build a field on, idiosyncratic enough to demand respect, and influential enough that every choice about it becomes a choice about what, exactly, is being switched on and off inside the living brain.

6.2 The Genetic If-Statement

Injecting a virus into a particular patch of tissue can feel wonderfully direct: the courier arrives at the address you chose, and cells nearby begin manufacturing an opsin. The trouble is that a neighborhood is not a household. Even a tiny volume of brain contains a mix of neurons with very different jobs, plus glial cells that support and insulate them. If light changes an animal's behavior after a regional injection, the scientist is left with a sharper question than "where?": *which kind of neuron was actually manipulated?*

The most common way to tighten that question is to add an "if-statement" to the genetic payload. In ordinary language, an if-statement is a rule like: *if* the user enters the correct password, *then* unlock the phone. Biology has its own passwords. Cells carry distinctive patterns of gene activity, and molecular biologists have learned to exploit those patterns so that an opsin turns on only when certain conditions are met.

Two ingredients do most of the work: promoters and recombinases.

A *promoter* is a stretch of DNA that helps decide whether a gene is transcribed into RNA. It is not an on/off switch in the electrical sense; it is more like a sign on a door that says "staff only," followed by the reality that some doors are guarded more strictly than others. Put a promoter in front of an opsin gene and you can bias expression toward neurons in general, or toward certain broad families of neurons. If the promoter is active mainly in excitatory neurons (the ones that tend to push their partners toward firing), then the opsin will mostly appear there. If it is active mainly in inhibitory neurons (often using the neurotransmitter GABA, which tends to restrain firing), then the opsin will mostly appear in those.

"Mostly" is doing important work. Promoters in viral vectors are compact, portable fragments of a larger regulatory landscape, and the full original gene may be controlled by far-flung enhancers and chromatin context that cannot be packed into a small payload. So a promoter can be wonderfully useful and still leaky at the edges. In a complex tissue, even a small amount of expression in an unintended population can become relevant, especially when optogenetics is powerful enough that a few strongly driven neurons can change a circuit's mood.

When the goal is sharper specificity, neuroscientists often reach for a second ingredient: a *recombinase*. Recombinases are enzymes that can cut and rejoin DNA at specific short sequences. The most widely used is Cre, which recognizes lox sites (hence "Cre/lox"). Another common one is Flp, which recognizes FRT sites ("Flp/FRT"). If promoters are door signs, recombinases are locksmiths that can permanently rearrange the lock.

Here is the key idea that turns recombinases into logic: the virus can be everywhere in the injected region, but the opsin stays silent unless the target cell contains the recombinase. Expression occurs only when recombinase is present in the target cell population, enabling cell-type specificity when paired with driver lines or recombinase-delivering vectors. A cell that lacks Cre can still be infected by the virus and still carry the DNA payload, but it will not be able to "unlock" it.

That separation—infection versus expression—is the heart of the genetic if-statement. It also helps explain why optogenetics can feel simultaneously precise and messy. The virus can spread in a cloud, yet the actual light-sensitive protein can appear in a more selective pattern, filtered by a second layer of rules inside cells.

The genetic vocabulary for "locked" and "unlocked" can sound ceremonial: "floxed," "Cre-dependent," "conditional," "driver line." Underneath is a simple trick. The opsin gene is placed in a configuration that cannot be read properly until Cre has acted. Sometimes the gene is inverted, like a line of text printed upside down. Sometimes a stop signal blocks transcription until it is excised. The arrangement is engineered so that a cell with Cre performs a one-time edit to the DNA, after which the opsin can be expressed normally and robustly. Because the edit is permanent at the DNA level, the cell's descendants (if the cell divides) would also carry the change. Neurons generally do not divide, but the permanence still matters: it means the cell does not need continuous Cre activity to keep the opsin on; it only needs Cre long enough to flip the switch.

Among the most popular designs is FLEX/DIO, short for *flip-excision / double-floxed inverted ORF*. The name is technical; the motivation is practical. FLEX/DIO is widely used to reduce leaky expression compared to some stop-cassette designs, especially in high-copy AAV contexts. AAV infections can deliver multiple copies of the cassette into a single cell, and any small tendency toward accidental expression can add up when you have many copies. FLEX/DIO uses pairs of recombination sites arranged so that Cre first flips the gene into the correct orientation and then removes extra sites to stabilize the new configuration. The result is a system that behaves more like an intentional gate and less like a creaky door.

If this sounds like software engineering, that resemblance is not accidental. Modern neuroscience has become comfortable thinking in layers of constraint. A promoter asks, "Where can this code run?"

Cre asks, "Is the condition satisfied?" FLEX/DIO stores the code backward until the condition is met. You can build remarkably selective expression not by finding a single perfect marker, but by stacking imperfect filters.

A concrete example makes the logic less abstract. Imagine trying to manipulate inhibitory interneurons in a structure involved in anxiety-like behavior. One approach is to use a promoter that tends to be active in inhibitory neurons, package the opsin behind that promoter, and inject locally. That may enrich expression in inhibitory neurons, but some excitatory neurons might still pick up expression, and some inhibitory subtypes might be missed. A Cre-based approach changes the game. If an animal has been engineered so that Cre is expressed primarily in inhibitory neurons (a so-called Cre-driver line), then a Cre-dependent AAV can be injected into the relevant area. Many cells will be infected, but only those that contain Cre will rearrange the cassette and produce the opsin.

The phrase "Cre-driver line" carries a hint of its own history. Cre/lox did not begin as a neuroscience tool; it was developed as a genetic system in bacteria and then adopted widely in mice for conditional genetics, allowing genes to be knocked out or activated only in particular tissues. Neuroscience benefited enormously from that existing toolbox. Once transgenic mouse lines were created to express Cre in particular neuronal classes, optogenetics could piggyback on them. AAV did not have to carry the Cre gene itself; it only had to carry a Cre-dependent opsin cassette. That division of labor is part of how researchers stay within AAV's packaging limits while still achieving strong, selective expression.

Promoters, however, did not become obsolete. They often appear again inside the viral cassette, even in Cre-dependent designs, to tune expression levels and bias expression toward neurons rather than glia. The common reality is layered gating: a promoter that sets the overall tone ("prefer neurons"), combined with recombinase logic that sharpens it ("only the neurons that meet condition X"). It is the biological equivalent of needing both a keycard for the building and a key for the office.

Intersectional targeting takes the logic one step further. If Cre is one condition, Flp can be another. Intersectional strategies (e.g., requiring both Cre and Flp) can narrow expression to cells defined by multiple features (marker + projection + location), but add operational complexity (multiple vectors, timing, titers). In plain terms, instead of "Cre *or* not," you can create "Cre *and* Flp." That can be decisive when a single marker is too broad.

Consider a common frustration: a gene marker may be shared by neurons across several areas, or by several subtypes within one area, only one of which is believed to drive the behavior of interest. An intersectional plan might look like this. First, define a broad class with Cre (say, inhibitory neurons). Next, define a projection pattern using a retrograde-capable vector that brings Flp (or Cre) back from a downstream target to the upstream neurons that connect to it. Finally, deliver a payload that requires both recombinases to activate. Only neurons that satisfy both conditions—belonging to the genetic class and sending axons to the chosen target—turn on the opsin. Nothing about this is mystical. It is a carefully arranged set of permissions.

These multi-vector strategies are where the "probabilistic" side of targeting becomes unavoidable. Each step has an efficiency: the fraction of relevant neurons that actually get infected, the fraction that express enough recombinase, the fraction that recombine the cassette. Each step also brings its own off-target possibilities. A retrograde vector may label unexpected afferents; a driver line may have low-level expression outside the desired class; the payload virus may spread farther than planned. Add more gates and you narrow the expression pattern, but you also increase the number of places the plan can fail quietly.

The quiet failures are often the most instructive. A common surprise is that Cre marks lineage rather than adult identity. Cre-driver lines can produce unintended recombination due to developmental expression, germline activity, or incomplete specificity; validating driver specificity via reporter crosses and careful breeding/genotyping strategies is standard practice. A neuron might express the marker gene transiently during development and then shut it off in adulthood. If Cre was active during that developmental window, it could permanently recombine the cassette, meaning the adult neuron now expresses the opsin even though it no longer expresses the marker. To a researcher who thinks of Cre as a label for adult inhibitory interneurons, this can look like a baffling betrayal. To a developmental biologist, it is exactly what one would expect from a permanent DNA edit driven by a transient signal.

Germline recombination is another kind of trap. If Cre is active in reproductive cells, recombination can occur in embryos in a way that makes the conditional system unconditional: the "locked" cassette becomes unlocked in every cell from the start. The resulting animal may look normal, and the genotype may still look superficially correct unless breeding and genotyping are handled carefully. The experimenter thinks they are doing a targeted manipulation; the biology has quietly turned it into a broad one.

Because of these pitfalls, responsible use of Cre lines often begins with a "reporter" cross: breeding the Cre-driver animal with another line that contains a silent fluorescent marker that turns on only after Cre recombination. If fluorescence appears only where it should, confidence increases. If it appears in unexpected places, the driver line is not discarded so much as re-understood. It may still be useful, but only for questions it can honestly answer.

AAV adds its own twist. AAV does not necessarily infect all neurons equally. Even within a Cre line that is perfectly specific, the actual opsin-expressing population may be a biased sample of the Cre-positive neurons—biased by vector tropism, injection details, and tissue properties. In other words, recombinase logic can be exact at the level of DNA editing, while the delivery that brings the cassette is uneven. This is one reason many optogenetic papers show expression maps and quantify spread: not as decoration, but as evidence of what population was realistically manipulated.

The most uncomfortable limitation is that specificity is rarely absolute. Promoters can be "quietly active" at low levels in unintended cells. Cre-dependent cassettes can exhibit a small amount of Cre-independent expression, especially when expression is pushed hard with strong regulatory elements or high viral titers. Recombinases themselves can sometimes have toxicity at very high levels, or cause recombination at pseudo-sites in the genome in extreme cases. These are not reasons to abandon the tools; they are reasons to treat targeting as something you measure, not something you assume.

A concrete behavioral example shows why small leaks can loom large. Suppose the scientific question concerns a sparse population of dopamine neurons implicated in reinforcement and relapse. The intended experiment is to activate only those dopamine neurons. If the targeting strategy accidentally turns on the opsin in a small number of neighboring non-dopamine neurons that project widely, the behavioral effect may reflect a mixture of influences. The effect could be real and reproducible while still being misattributed. In a system as interconnected as the brain, a few misplaced switches can be enough to create a compelling story that is not quite the right one.

That is why genetic targeting often feels less like stamping "VIP" on a select group of neurons and more like writing a careful contract.

The contract can specify location (where the virus goes), a general category (which promoter can speak), and a condition (which recombinase must be present). It can even specify two conditions at once. But the contract is signed by living cells, not by a computer, and living cells interpret instructions with variability.

There is a certain humility built into the best uses of these tools. The aim is not to pretend that a promoter or a Cre line defines a neuron category perfectly. The aim is to build a manipulation that is *biased enough* toward the intended population that a causal claim becomes plausible, and then to gather enough evidence—expression patterns, physiology, controls—that the claim does not rest on wishful thinking.

When it works, the result feels almost uncanny: light delivered to a small region shifts an animal's behavior, and the shift can be traced to a recognizable neuronal class. When it fails, it often fails in enlightening ways. A promoter is weaker than expected in one area, hinting at unappreciated gene regulation. A Cre line labels more broadly than anticipated, reminding researchers that "cell identity" is not a single label but a history. Even the leaks teach an important lesson: the brain does not come pre-partitioned into the categories scientists would like, so the most honest targeting strategies are those that treat specificity as a probability distribution rather than a binary badge.

The genetic if-statement, then, is less a magic wand than a disciplined way of asking: *Under what molecular conditions will this cell be allowed to respond to light?* The power comes from being explicit about the conditions, and from remembering that in biology, every condition is implemented by parts that are sturdy, imperfect, and profoundly real.

6.3 Needles, Coordinates, and a Brain Wrapped in Fog

After the genetics comes the geometry. Promoters and recombinases can narrow expression to a chosen neuronal population, but they cannot place that population in the skull. Someone still has to guide a needle into living tissue, deliver a few hundred nanoliters of viral solution, and later position an optical fiber so light reaches the same volume. It is a feat of scale: the experimental plan is drawn in millimeters, the relevant nuclei can be smaller than a lentil, and the margin for error is often a few tenths of a millimeter.

Neuroscientists do not do this freehand. They use *stereotaxic surgery*, a method that turns the head into a three-dimensional coordinate system. The animal's skull is held in a rigid frame. Adjustable arms and micrometer drives allow the surgeon to move a needle in three axes: left—right, forward—back, and up—down. If it sounds like machine tooling, that is the right mental image. The astonishing part is that this mechanical precision is being used to find soft targets inside a variable biological object.

The map for these coordinates is an *atlas*: a set of annotated brain slices with names, boundaries, and numbers. Most rodent atlases are referenced to skull landmarks, especially bregma, the point where two skull sutures intersect. In principle it is wonderfully simple: find bregma, set it as zero, and then move to the desired anterior—posterior and medial—lateral coordinates. Finally, lower the needle to the dorsal—ventral depth that matches the atlas.

In practice, stereotaxic targeting relies on atlases (commonly bregma-referenced coordinate systems), but published coordinates can disagree across atlas editions and across papers; atlases are best treated as starting approximations. That sentence carries a lot of lived experience. Atlases are built from averaged brains and carefully prepared slices, but real animals are not averages. Brains are slightly stretched or compressed by fixation. The skull can be angled a touch differently in the frame. The animal can be a little younger or older than the atlas standard. Even within a strain of laboratory mice or rats, the size of the brain and the relative position of structures can drift enough to turn a good plan into a near miss.

A concrete story circulates in many labs, usually told with a weary affection for the atlas rather than contempt. Two groups try to target the same tiny nucleus using coordinates published in reputable papers. Both "followed the atlas." One group uses an older edition of a standard atlas; another uses a newer edition or a different reference. The nucleus is only a few hundred microns long in the anterior—posterior direction. The published coordinates are shifted by a few tenths of a millimeter, perhaps because the atlas plates were aligned slightly differently, or because the defined border of the nucleus changed between editions. Both groups hit something real, both see viral expression, both implant fibers, and both can produce behavioral effects with light. Months later, when they compare histology, it becomes clear that one group has been working one neighborhood over. Their "same target" experiments are not contradictory so much as aimed at adjacent pieces of circuitry. The disagreement is not resolved by arguing about who read the atlas correctly; it is resolved by accepting what the atlas is: a reference, not a guarantee.

Even with a good atlas, the practical precision of stereotaxic injections is limited—especially for small nuclei—and may require multiple injections for adequate coverage of larger regions. That limitation is not a confession of incompetence. It reflects the realities of delivering fluid into tissue and the realities of anchoring coordinates to an external skull. A nucleus can be wrapped around another structure, or have a curved shape that makes it hard to "fill" with a single injection. In those cases, a lab may choose two or three injection sites, each small, rather than one larger bolus that risks spreading into neighbors. The goal becomes a careful compromise: enough expression in the intended area to manipulate it, without turning the manipulation into a blunt instrument.

The surgeon's craft begins before the needle ever touches brain. The skull must be leveled so that the coordinate axes match the atlas. Small differences in pitch and roll can translate into large depth errors for deep targets. The drill hole must be placed accurately. The needle or glass pipette must be thin enough to reduce damage, yet sturdy enough not to bend. Each of these choices makes the difference between a clean track and a messy one, and a clean track is not only about animal welfare; it affects where the virus goes.

Once the needle enters tissue, the atlas becomes less helpful and the physics becomes more important. The injection itself is a controlled injury. The needle creates a channel. The injected solution increases local pressure. The tissue can elastically deform and then partially rebound when the needle is withdrawn. Backflow can carry solution up the track, leaving a streak of expression above the target. A slow injection rate can reduce backflow but extends the time the tissue is displaced. A fast injection can cause reflux and a wider, more irregular spread. A larger volume can reach more neurons but also creates more disruption.

This is why many labs empirically adjust atlas-derived coordinates until histology verifies correct targeting; this is an iterative calibration loop, not a one-shot lookup. A new project often begins with what is essentially a rehearsal: inject a fluorescent reporter into the intended area, wait for expression, slice the brain, and see where the fluorescence landed. If it is consistently too shallow, the next cohort gets a slightly deeper dorsal—ventral coordinate. If it is too lateral, the medial—lateral value is nudged. Sometimes the adjustment is tiny, on the order of 0.1—0.2 mm, which sounds trivial until you remember that neighboring nuclei can be separated by that amount. The calibration repeats until the lab can reliably hit the intended structure in that strain, at that age, with that surgical setup, using that needle and injection protocol.

Calibration is also social knowledge. A senior technician may be able to look at a skull, or feel the resistance of the needle as it passes through tissue layers, and predict whether the placement is likely to be accurate. That skill is hard to convey in a methods paragraph. It is one reason labs guard experienced surgical expertise as carefully as they guard expensive lasers.

The fiber optic adds its own layer of geometry. For many experiments, the fiber is implanted into or just above the area where the opsin is expressed. The fiber must be secured to the skull so it stays stable for days or weeks, surviving normal movement. Deep targets often require penetrating fibers or cannulas, creating trade-offs: access versus tissue disruption and potential activation of passing fibers or neighboring structures. A fiber has a finite diameter, so it displaces tissue. The path to the target may pass through other structures, creating the possibility of unintended damage or inflammation along the route. In a brain, passing fibers are not just anatomical clutter; they are living axons carrying information. Disrupting them can change behavior even without light. Then, when light is delivered, those same axons may contain opsin-expressing terminals from elsewhere, adding another layer of potential off-target influence.

So far, the uncertainty has been mechanical and anatomical. There is another uncertainty that feels almost philosophical: light does not behave in brain the way it behaves in air.

The brain is not transparent. It is a scattering medium, with cells, membranes, and myelin constantly bending and diffusing photons.

As soon as light exits the fiber tip, it begins to spread and weaken. Some of that spreading is straightforward geometry, like the widening of a flashlight beam. Much of it is scattering and absorption. The net effect is that the "illuminated volume" is broader and blurrier than the intuitive image of a narrow beam. The fog is also optical: scattering and absorption make light distribution broader than the intended beam, and simplistic isotropic scattering assumptions can be invalid at optogenetic length scales; fiber geometry and launch parameters matter. In other words, even if one had a perfect map of where photons go in a uniform gel, the real brain can bend those photons in ways that depend on local microstructure.

The details of the optical setup shape this blur. The numerical aperture of the fiber (a measure of how widely it emits light), the wavelength used, the power at the fiber tip, and the coupling between laser and fiber all influence how far light penetrates and how wide it spreads. Blue light, commonly used for certain opsins, scatters more strongly than red light, so its effective reach in tissue is typically shorter. Red-shifted opsins were developed partly to take advantage of that physics: longer wavelengths generally penetrate further, though biology and opsin kinetics impose their own constraints. None of this needs to be calculated in a reader's head to appreciate the main consequence: "fiber above the target" does not mean "only the target receives light."

This optical fog interacts with the expression fog from viral spread. The virus may create a gradient of opsin expression, and the light creates a gradient of illumination, and the product of those gradients determines which neurons are actually perturbed. A neuron that expresses a lot of opsin but sits at the dim edge of the light field may be modulated modestly. A neuron that expresses little opsin but sits near the fiber tip may still be driven strongly. Neurons do not politely announce where they are on this map; the map is inferred from physics, histology, and the behavioral outcome.

A dramatic illustration comes from experiments that target small nuclei involved in arousal or stress. These regions can be nestled among other nuclei with overlapping functions. A researcher may aim for a tiny cluster and place the fiber just above it to minimize damage. The light, however, does not stop at the nucleus boundary drawn in the atlas. It spreads into the neighboring structure. If the neighboring structure expresses the opsin (perhaps due to viral spread, or because the genetic strategy labels neurons there too), the behavioral effect could arise from a mixture of influences. The resulting conclusion may still be true at a coarse level—the general area participates in the behavior—but it becomes risky to claim that a single nucleus is the causal node.

The surgeon's response to this is not despair; it is strategy. For some questions, broad modulation is acceptable, even desired. For others, the setup is designed to minimize spread: smaller injection volumes, more restricted viral serotypes, lower light power, fibers with smaller cores, careful placement deeper into the target. Sometimes the answer is to move away from illuminating cell bodies and instead illuminate axon terminals in a downstream area, using anatomy as a filter: only neurons that project to the illuminated site can influence the effect. That strategy brings its own complications—axons of passage, local collateral branches, and the possibility of antidromic activation (signals traveling back to the cell body)—but it shows how targeting is often a conversation between anatomy and physics rather than a single trick.

There is also a moral tension that runs through these decisions, and it is worth naming plainly: every added layer of precision tends to increase invasiveness. A deeper fiber gives better access but creates more damage. A thinner needle reduces disruption but can clog and make injections less reliable. A smaller injection volume reduces spread but may not transduce enough neurons to produce a measurable behavioral change. Optogenetics encourages crisp causal claims, yet the physical steps that enable those claims inevitably perturb tissue. The best experiments are not those that pretend this tension does not exist; they are those that design around it and test for the artifacts it can create.

A final piece of the story is that the brain, even within one animal, is not fixed. Tissue can swell after surgery. Scar tissue can form around an implant. The fiber can shift slightly over time. The skull itself can change subtly as the animal grows. These shifts are usually small, but when the targets are small, "small" is not always reassuring.

Stereotaxic surgery, then, is often described with the language of coordinates and apparatus, but experienced practitioners talk about it as a craft. It is a craft built on measurement and repetition, on accepting that atlases are starting approximations, on treating published coordinates as hypotheses, and on learning how a particular lab's hands, needles, pumps, and animals behave. The reward is not a perfectly pinpointed manipulation. The reward is a manipulation whose uncertainties are understood well enough that a scientist can look at an effect—a freezing mouse, a lever press, a startled pause—and know which anatomical claims are justified and which are reaching beyond what the needle and the light can honestly deliver.

In the end, the coordinate system does something more than help a needle find a target. It forces a habit of mind: to translate a clean, named structure into a volume of living tissue that has edges you cannot see during surgery, that moves and yields, and that sits inside an optical fog. Working within that fog is not a failure of precision; it is the everyday condition of doing causal neuroscience in a real brain.

6.4 A Target Is Real Only After You Check It

Plans are clean. A coordinate is three numbers. A genetic strategy can be written as a tidy sentence: “Cre-dependent opsin in inhibitory neurons.” A fiber optic has a specified length and a polished tip. On paper, the experiment already sounds like it has *targeted.*

In practice, targeting is not a promise you make at the start. It is something you earn at the end, by asking questions that feel almost rude in their insistence: Did the opsin end up where it was supposed to? Did it land in the intended cells rather than their neighbors? Is the fiber actually where you think it is? When light was delivered, did the manipulated neurons respond the way the tool is supposed to make them respond? And could a more mundane explanation—heat, damage, stress, distraction—account for what looks like a clean causal effect?

The way optogenetics matured as a field is partly the story of how these questions became non-negotiable. Early enthusiasm sometimes encouraged a kind of false precision: the comforting idea that a coordinate plus a virus plus a laser equals a defined manipulation. Over time, laboratories learned that the true unit of evidence is not the surgical plan but the verified anatomy and verified physiology of each animal.

The first and most familiar layer of checking is *post hoc histology.* “Post hoc” simply means after the fact: after the behavior, after the light sessions, after the data have been collected. The animal is perfused, the brain is fixed, sliced into thin sections, and examined under a microscope. The viral payload usually includes a fluorescent reporter—a protein that glows under the right illumination—either fused to the opsin or expressed alongside it. That fluorescence turns the brain into a map of where the opsin is likely to be.

Post hoc histology commonly verifies (a) opsin/reporter expression location and spread and (b) optical fiber/cannula placement relative to the intended structure; many papers explicitly report these checks and may exclude mistargeted subjects. The phrase "exclude" can sting, especially for a reader imagining a one-off experiment with a heroic mouse and a single, definitive dataset. In reality, exclusion is often the most honest outcome. If the fiber track sits outside the intended area, or if expression sprawls into an adjacent nucleus in a way that makes the causal claim ambiguous, the animal is not a "failed mouse." It is a reminder that targeting is an empirical outcome with a failure rate, and that the field's credibility depends on not hiding that fact.

Histology also answers a question that is easy to forget in the excitement of a behavioral result: *is there enough expression to plausibly do anything?* An opsin experiment can fail quietly if expression is weak. A negative behavioral result might reflect biology ("this circuit is not involved") or logistics ("the switch was barely installed"). Without histology, those two explanations can look identical. With histology, they separate: you can see whether the experiment truly perturbed the intended tissue.

A deeper histological check goes beyond "where is the fluorescence?" and asks "*what* is fluorescing?" This is where immunostaining and marker colocalization come in. Researchers can stain for proteins that label broad neuronal categories—for example, markers of inhibitory neurons—and see whether the opsin signal overlaps. This is a way of checking whether a promoter or a Cre-driver line behaved as hoped in that particular anatomical context. Because recombinase logic can be undermined by developmental recombination or imperfect specificity, this step is often the difference between a confident claim and an attractive guess.

Verification is necessary because multiple layers can fail: coordinate mismatch, injection spread/tissue damage, off-target recombinase activity, and optical scattering that expands the effective stimulated volume. Histology catches several of these at once. Coordinate mismatch shows up as expression centered in the wrong place. Injection spread shows up as a halo or smear that crosses boundaries drawn in the atlas. Tissue damage can be visible as a tract, a cavity, or gliosis around the implant. But histology has limits too: fluorescence is not the same as functional opsin current, and a fiber track does not tell you exactly how light spread through the tissue in the living animal.

That brings us to the second layer of checking: *did the cells actually respond to light?* This is functional validation, and it can be done in several ways.

One approach is electrophysiology in brain slices. After expression has developed, a subset of animals can be sacrificed for acute slice recordings. A neuron in the targeted area is patched with a glass pipette, its electrical activity is measured, and light is delivered through the microscope. If the opsin is excitatory, light should evoke currents and often spikes with the expected timing. If the opsin is inhibitory, light should hyperpolarize or suppress firing. This is a direct confirmation that the tool is present at a sufficient level and is doing what it is supposed to do in the relevant cellular environment.

In vivo electrophysiology can serve the same purpose, though it is technically more demanding. A lab can record from neurons near the fiber tip while delivering light, checking for light-locked changes in spiking. This is particularly valuable because it tests the full preparation: the opsin, the fiber placement, and the optical power in living tissue. It also exposes an uncomfortable truth: light can influence recorded signals in indirect ways, for example by driving nearby neurons that then synaptically affect the recorded neuron. That indirectness is not necessarily a flaw—it may reflect the very circuit interaction one hopes to study—but it means that "light changes spiking" is not automatically the same as "light directly controls these neurons." The details matter.

Calcium imaging provides a third perspective. Instead of recording electrical spikes directly, calcium imaging measures a fluorescent signal that tracks activity-related calcium influx. It is slower than electrophysiology but can report activity in many neurons at once. When paired with optogenetics, it can show whether light changes the activity pattern of the targeted population, and whether those changes remain confined to it or ripple outward. Imaging is not a replacement for electrophysiology, but it often functions as a reality check: a way to confirm that the manipulation actually moved the circuit.

None of these functional checks is purely technical. They are ethical checks on interpretation. Optogenetics makes it easy to confuse intention with effect: "I shone light on the fear circuit, therefore I altered fear." Functional validation forces a more precise statement: "I shone light, and neurons with opsin in this location produced currents of this size, with this timing, at this power." It is less poetic, but it is the backbone of a credible causal claim.

A third layer of checking concerns *controls*, and controls in optogenetics are often where the most interesting skepticism lives.

The simplest control is a control group that underwent the same surgery and fiber implantation but received a virus that expresses only a fluorescent reporter, not an opsin. These animals experience the same tissue disruption, the same weeks of viral expression, the same handling, and the same light delivery, but their neurons should not be light-sensitive. If light changes behavior in the control group, the experiment has discovered an artifact: perhaps the light is heating tissue, or perhaps the cable and commutator alter the animal's movement in a way that interacts with the task.

Control wavelengths are another tactic. Many opsins respond strongly to one wavelength and weakly to another. Delivering a wavelength that should not activate the opsin (or activating it at a power too low to be effective) can reveal whether the mere presence of light in the brain changes behavior. This is not a trivial concern. Light can heat tissue, and heating artifacts are real: a small temperature increase can influence neuronal firing, synaptic release, and blood flow. The worry is not that every optogenetic experiment is secretly a heat experiment; the worry is that a subtle thermal effect could be misread as a circuit-specific manipulation, especially when the behavioral readout is sensitive and the animals are stressed.

Sham stimulation is a related control: connecting the animal to the optical cable and running the same stimulation protocol with the laser shuttered. This controls for the sensation of cable attachment, the sound of equipment, and the timing cues that the animal might learn. In behavioral neuroscience, animals are excellent at learning about the experimenter's world. If light pulses always coincide with a particular trial type or time window, an animal may change its behavior based on that cue alone, independent of any neural manipulation. Controls are a way to ensure the "cause" is neural, not theatrical.

At this point, it may sound as if verification is a long list of chores. The deeper reason it is compelling is that it repeatedly rescues scientists from being fooled by their own competence. A well-run surgery can still miss. A well-designed genetic strategy can still express outside the intended population. A beautiful behavioral effect can still come from a neighboring structure. Verification is the antidote to the seductive feeling of false precision.

A concrete mini-case shows how this plays out. Imagine a lab studying avoidance behavior and targeting a small nucleus thought to encode threat prediction. They run the experiment and see a striking result: when light activates the opsin, animals abruptly stop entering an open arm of an elevated maze. The effect is reliable and large. It is tempting to declare victory and attribute the behavior to the target nucleus.

Then histology is done. The fluorescence is indeed bright and localized—but it is centered just beside the intended nucleus, in a neighboring region with different connections. The fiber track sits slightly posterior to the planned coordinate. The outcome is not that the experiment "failed." The outcome is that the story changes. The causal claim must now be anchored to the verified anatomy, not to the intended target. In some cases, this leads to a new insight: the neighboring region may be more important than the original candidate. In other cases, it simply prevents a wrong conclusion from becoming a citation that misdirects future work.

Retrograde and broadly tropic vectors can label multiple populations (including cells at the injection site and many afferents), so intersectional design and verification are critical to claim 'defined circuit' manipulation. Retrograde strategies, in particular, are powerful but easy to misinterpret. Injecting a retrograde-capable AAV into a target area can lead to expression in many upstream populations that project there. That can be exactly what you want—a way to manipulate a pathway defined by its projection—but it also creates a bookkeeping problem: which upstream area is actually driving the behavioral change? If several afferent populations are labeled, light delivered at one site might influence terminals from multiple sources. Intersectional logic can narrow the expression, but it also increases the number of moving parts that must be checked. Histology becomes less like a ceremonial figure and more like an audit.

Verification also responds to a sociological problem in science: coordinates travel too easily. A paper publishes "AP -1.4, ML +0.3, DV -4.8," and another lab adopts it. Yet coordinate discrepancies across references and individual-animal variability motivate documenting both planned coordinates and verified anatomy rather than relying on coordinates alone. Reporting verified placements is a way of keeping the scientific conversation honest. It acknowledges that the atlas is a starting point and that the true experiment occurred in a particular animal's brain, with particular deviations.

Many laboratories therefore develop a style of documentation that resembles a checklist. Expression: present, absent, too broad, too weak. Fiber placement: correct, shallow, lateral, posterior. Damage: minimal, substantial, along the track. Included animals: those that meet pre-defined histological criteria. Excluded animals: those that do not. This approach can feel ruthless, but it protects the meaning of the manipulation. Without it, the term "targeted" becomes a compliment rather than a description.

There is a final, subtle form of checking that happens not under a microscope but in the logic of the experiment itself: building in ways for the result to prove you wrong. If stimulating the target produces a behavioral change, does inhibiting it produce the opposite change? If the effect appears only during a specific task phase, does it vanish when stimulation is shifted in time? If the effect is claimed to be specific to one neuronal population, does it vanish in animals lacking the recombinase needed for expression? These logical controls are not substitutes for histology and physiology; they are additional constraints that make an artifact less likely.

A target becomes real when these layers line up: when the anatomy is verified, when the physiology shows that light actually perturbs the intended neurons, when the controls show that light alone does not explain the behavior, and when the behavioral logic is consistent with a circuit mechanism rather than a generic disturbance. The reward is not perfection. The reward is a manipulation that can withstand the most valuable kind of attention in science: someone else trying very hard to find an easier explanation.

Chapter 7

Wires of Glass: How Light Reaches Deep Circuits

Optogenetics sounds almost magical until one asks the awkward practical question: how does light actually reach a neuron buried millimeters inside the brain? Following that question leads into a world of wavelengths, implanted fibers, scattering tissue, and increasingly clever tricks—from tethered ferrules to holographic pulse patterns—that reveal why every flash of light is also an engineering compromise.

7.1 Color as Tactic: Blue, Red, and the Physics of a Decision

A beam of light looks clean and decisive in air: a tidy cone, a clear destination. The same beam entering brain tissue becomes something more like a rumor—scattered, muffled, and reshaped by whatever it meets along the way. Optogenetics lives inside that mismatch. The molecules may be exquisitely engineered, and the genetics may be surgically selective, yet the light itself must still negotiate an environment that was never built to be transparent.

The color of that light is the first tactical choice. Not because blue looks more "high-tech" or red feels more "gentle", but because color governs how quickly photons get lost and where their energy ends up. In practice, color sets the balance among three hard constraints: how deep you can effectively illuminate, how much power you must deliver to recruit enough opsin molecules, and how much unwanted heating and photochemistry you risk while doing it.

Brain tissue blocks light in two main ways: absorption and scattering. Absorption is the straightforward one: photons are captured by molecules and converted into other forms of energy, often heat. Scattering is sneakier: photons are not destroyed, but bounced around, changing direction again and again until the original beam no longer resembles a beam. When someone says "light doesn't penetrate well," it is often scattering, not absorption, doing most of the work. The brain is full of tiny refractive-index mismatches—cell membranes, organelles, myelin, extracellular matrix—so even a beautifully collimated beam quickly becomes a fuzzy cloud.

Color matters because both absorption and scattering depend strongly on wavelength. The most general rule is that longer wavelengths tend to scatter less in tissue. Blue photons encounter more frequent deflections; red photons, on average, travel straighter for longer. The effect is not subtle. A small change in wavelength can shift scattering enough to change whether illumination stays near the fiber tip or spreads into an unruly volume. For optogenetics, that translates into an experimental reality: when you use blue light, you are often paying an unavoidable "fog tax". You may still control neurons with millisecond precision, but you usually do so within a smaller, shallower neighborhood, or by using more power to compensate for what is lost on the way.

Absorption adds its own geography. Brain tissue contains chromophores—molecules that soak up certain colors. Hemoglobin, in blood, is a powerful player. It absorbs strongly in the blue and green, less in the red. That means that a vascular patch—capillaries you cannot see, blood volume that changes with physiology—can become a moving target for your light budget. Blue light is not simply dimmed; it can be locally swallowed by blood, converting optical energy into heat precisely where tissue is most sensitive. Red light is not immune to absorption, but it enjoys a friendlier corridor through the spectral landscape. Water, meanwhile, becomes more absorptive at still longer wavelengths, eventually turning "more red" into "more heating" rather than "more depth". Optogenetics thus inherits an awkward truth from photomedicine: there is a rough window where tissue is relatively permissive, and outside that window the body starts behaving like a sponge.

This is why the early triumph of blue-light optogenetics carried, embedded within it, an engineering ceiling. Channelrhodopsin-2, the workhorse that helped launch modern optogenetics, is driven by blue light. It delivered a kind of control neuroscientists had wanted for decades: not merely exciting or silencing neurons, but doing so on command, in patterns that could match the pace of spikes. It was a molecular switch that respected the nervous system's native tempo. Yet blue's advantages in molecular compatibility did not magically change the optical properties of mammalian brain.

A concrete example makes the constraint vivid. In 2007, Adamantidis and colleagues, working with Feng Zhang and collaborators, used optogenetic stimulation of hypocretin (orexin) neurons to probe the circuitry of waking. Hypocretin cells sit deep, in the hypothalamus—a location that is physiologically central and optically inconvenient. The experiment's conceptual heart was simple: if these neurons are made to fire in a wake-like pattern, does the animal wake? But the success of the idea depended on something more prosaic: could enough light be delivered to enough opsin molecules, deep in a living brain, without cooking tissue or losing the animal's normal behavior under a tangle of hardware? Blue light could do the job, but it required careful compromises—fiber placement close to the target, power high enough to matter, pulse patterns chosen not only for physiology but also for what the tissue could tolerate. The biology and the optics were not separate problems; they were a single design.

The first wave of optogenetics therefore taught an unintuitive lesson. Even if you know exactly which neurons carry your opsin, you do not automatically know which neurons your light will actually influence. Blue light tends to create a more confined effective zone, not because it is "sharper," but because it gets scattered and absorbed more strongly. If you want to reach a nucleus two millimeters below the surface, blue will not politely travel there just because you aim it. You either bring the light source closer—often by implanting a fiber—or you accept that the required surface illumination would be so intense as to be impractical or damaging.

Red-shifted opsins entered this story as a practical liberation. By moving the opsin's sensitivity toward longer wavelengths—into orange or red—researchers could exploit tissue's comparatively better transmission there. This was not a cosmetic tweak. It changed what kinds of experiments could be done without heroic light powers, and it softened a dilemma that had shadowed the field: to get enough light to deep structures, you risk delivering too much light to everything in between.

When an opsin is driven by red light, several good things happen at once. Less scattering means more of the light's "intent" survives as it travels. Lower hemoglobin absorption means fewer photons are taken hostage by blood on the way in, and fewer are converted to heat in the most vascular microenvironments. In many situations, this lets the experimenter use lower power at the source for the same biological effect at the target. The animal benefits: less heat, less photochemical stress, and fewer reasons for tissue to react in ways that masquerade as neural causation.

But red-shifted tools do not grant free wishes. They swap one set of constraints for another. Opsins are proteins with kinetics—opening and closing speeds, recovery times, desensitization behaviors. Some red-shifted variants, especially early ones, were not as fast or as light-sensitive as the blue classics, or they demanded higher photon flux to produce the same current. Light that travels deeper can still be too weak at the destination if the opsin is sluggish or dim in its response. There is also a more subtle trade-off: if you compensate for weaker opsin performance by increasing intensity, you may lose the heating advantage you hoped to gain. The color helped, but it did not repeal thermodynamics.

Temporal precision can also be part of the trade. Millisecond timing is optogenetics' signature promise, but not every opsin treats time equally. Fast opsins can follow high-frequency stimulation, letting researchers mimic spike trains that resemble natural firing patterns. Slower opsins act more like a sustained push than a crisp tap, which may be adequate—or even preferable—when the question is about state changes rather than precise spike timing. If you are probing whether a circuit can *enter* a fear state, a gentle sustained bias might be sufficient. If you are probing how a hippocampal rhythm shapes memory encoding on the scale of tens of milliseconds, sluggishness can blur the very phenomenon you are trying to test.

This is where step-function variants—a family of opsins engineered to remain active longer after a brief light pulse—become an intriguing counterpoint. Their appeal is not speed but efficiency. Instead of constantly hammering tissue with light to keep channels open, a short flash can latch the protein into an active state, and another flash can switch it back. The strategy is conceptually elegant: write a persistent change with minimal illumination. In a brain where heating and scattering impose costs, "less light for longer effect" can be a serious advantage. Yet step-function behavior changes the meaning of stimulation. A neuron held in an altered excitability state for seconds is not being "paced" in the natural language of spikes; it is being biased. That can be powerful for testing whether a population's activity is sufficient to produce a behavioral state, but it can also complicate interpretations about timing, coding, and synchrony. The color decision and the kinetic decision are entangled.

The deeper point is that wavelength is never chosen in isolation. It is chosen alongside geometry (how close the light source is to the target), expression levels (how many opsin molecules each neuron carries), and the biological question. A circuit question that sounds purely conceptual—Is this pathway necessary for reward learning? Can these neurons trigger a panic-like response?—turns, in the lab, into a budgeting exercise: photons in, biological effect out, confounds minimized.

Heating is the confound that makes this budgeting emotionally real. It is easy to imagine light as "clean," especially compared with drugs that diffuse widely and linger. But light is energy, and tissue has nowhere to send that energy except into molecular motion. Small temperature rises can change neural activity directly: ion channels shift their properties, synapses release transmitter with altered probability, blood vessels dilate, glia respond. Even before you reach the dramatic threshold of damage, you can cross into a zone where temperature itself nudges neural firing. That creates a nightmare scenario: you deliver light to test whether a circuit causes a behavior, and the tissue's thermal response becomes a silent partner in the result. The risk is highest when illumination is continuous, high power, or delivered through wavelengths and conditions that increase absorption. Red light often reduces that risk, but it does not eliminate it; it merely changes the slope of the trade-off.

This is why experienced optogeneticists talk about "power at the target," not "power at the laser." Between the source and the opsin sits a whole chain of losses and distortions. Even before light enters the brain, coupling efficiency—how well the beam enters the fiber, how cleanly connectors mate—can waste photons. Once in tissue, scattering and absorption immediately start reshaping the distribution. A milliwatt quoted at the source is a promise that may arrive partially broken. Color affects how badly the promise frays.

Color also influences the shape of what you perturb. Blue light, because it scatters more, can spill laterally near the source, creating a broader region of low-to-moderate illumination while failing to maintain intensity at depth. Red light tends to preserve directionality better over the first stretch, allowing relatively more power to reach deeper layers for the same source intensity. Neither is "more precise" in any absolute sense; the tissue's scatter turns precision into a negotiation. But the negotiation's terms change with wavelength.

There is a final, quietly profound consequence of going red: it helps optogenetics coexist with the technologies used to *watch* the brain. Many fluorescent indicators and imaging methods occupy the blue-green part of the spectrum. If your control light overlaps too much with your readout light, you can accidentally stimulate while you are trying to measure, or swamp your detectors with glare. Shifting stimulation to orange-red can create spectral breathing room, letting researchers monitor activity while manipulating it, each in its own optical lane. This is not merely a convenience. It changes what questions feel answerable, because it becomes easier to link a controlled perturbation to the living dynamics it produces.

The result is a view of color that is closer to strategy than style. Blue remains indispensable: many of the fastest, most robust excitatory opsins still live there, and in superficial brain regions the scattering penalty can be manageable. Red expands the playable map, letting experiments reach deeper with less collateral energy, and letting control coexist more gracefully with optical readouts. Neither color is "better." Each is a particular bargain with tissue: an agreement about how photons will be spent, where they will be lost, and what kinds of neural language—spike-precise or state-biasing—you can afford to speak.

A neuroscientist choosing a wavelength is therefore doing something like choosing an instrument for a difficult concert hall. The notes on the page may be the same, but the room changes everything: what carries, what echoes, what overheats, what blurs. In the brain, color is how you decide which parts of the room you are willing to fight, and which parts you will let the physics handle for you.

7.2 A Lantern on a Leash: Delivering Light to Moving Animals

Turning neurons on and off in a dish is a satisfying kind of magic: a flash, a spike; darkness, silence. The animal, though, refuses to be a dish. It sniffs, grooms, rears, runs, sleeps, startles, forgets you are there. If light is going to become a serious tool for understanding memory, fear, or craving, it has to arrive in the brain while the rest of the body carries on as if nothing unusual is happening. That requirement quietly reshaped optogenetics. The field did not only invent molecules; it learned how to route light through a moving world.

The basic idea sounds simple: bring photons from a source on the bench into a tiny target deep inside the head. The actual implementation is a chain of parts, each designed to solve an unglamorous problem. A laser or LED must deliver stable power; the beam must be coupled into a flexible cable; the cable must tolerate tugging and twisting; a connector must let you attach and detach animals without breaking the system; an implanted light guide must sit in the correct brain location for weeks; the entire assembly must survive cleaning, chewing, cage life, and the animal's normal restlessness. Most failures are not dramatic. They are a slow drift in coupling efficiency, a connector that picks up dust, a cable that develops a kink, an implant that loosens a fraction of a millimeter, or a patch cord that becomes just heavy enough to change how a mouse turns its head.

Start with the light source. LEDs are compact, relatively inexpensive, and can be switched quickly. Lasers provide high brightness and a clean, well-defined spectrum, and they are often used when power must be pushed through long cabling or when particular wavelengths are needed. Either way, the source is rarely the limiting factor. The difficult part is preserving a predictable fraction of that light all the way to the brain. A lab can own a powerful laser and still deliver an anemic trickle to the animal if the chain is poorly maintained.

The next step is coupling: getting the light into a flexible waveguide, usually a glass strand built to confine light by internal reflection. The coupling stage is where a surprising amount can go wrong. Small misalignments waste power. Dust on a connector face scatters light and can even burn the connector over time, leaving a permanent scar that further reduces throughput. Labs that do large numbers of behavioral optogenetics experiments often develop a kind of folk discipline around these tiny surfaces: cleaning rituals, inspection under a microscope, careful storage caps, rules about not touching polished ends. It sounds fussy until you realize that a behavioral experiment may hinge on a modest change in delivered intensity: the same pulse train that triggers a state change one day may do nothing the next if coupling quietly degrades.

From there light travels through a patch cable. This is where physics meets animal behavior in the bluntest way: the cable has weight, stiffness, and a tendency to twist. A mouse does not know it is attached to a delicate instrument. It moves in quick, looping paths. It turns its head repeatedly while grooming. It can spin while exploring a new corner. Without a way to release torsion, the cable becomes a spring. Tension rises, the animal feels it, and behavior changes for reasons that have nothing to do with neurons. The solution is the rotary commutator: a device that lets the cable rotate freely so the mouse can turn without winding itself into a tethered knot. In a successful setup, the commutator fades from the animal's experience. The experimenter, however, never forgets it is there, because any added friction or intermittent contact becomes an invisible variable.

Near the head, the chain reaches a connector and ferrule. A ferrule is a small, rigid sleeve—often ceramic or metal—that holds the end of the waveguide with a precisely polished face. It is the interface that makes chronic experiments practical. You can repeatedly connect an animal to the patch cable, run a session, disconnect it, and return it to its cage without redoing surgery or handling bare glass. Dennis Sparta and colleagues (2011) helped standardize how many labs build these implant assemblies for long-term use: how to seat the fiber in the ferrule, how to polish it so light exits cleanly, how to reinforce it with epoxy and strain relief so it survives weeks of handling. Standardization sounds bureaucratic until you have lived through the alternative, where every graduate student invents a slightly new connector style and no two animals have comparable light delivery.

Then comes the implanted portion: the segment that must sit stably in tissue. For deep targets, light cannot simply shine through the skull and cortex; it must be delivered close to the site of interest. This necessity pushed optogenetics toward surgical implantation early. Deep targets in freely moving mammals motivated implanted waveguides as early as 2007, enabling direct illumination of deep nuclei during behavior. That step was as conceptually important as any genetic trick. It meant optogenetics could leave the microscope stage and enter the arena of decisions, habits, and emotional learning.

Implant placement is usually performed with a stereotaxic frame: a mechanical system that holds the head in a fixed geometry while the surgeon navigates using a brain atlas. The language is millimeters and angles. A tiny error can place the tip above, below, or beside the intended circuit. Yet even perfect coordinates do not guarantee perfect targeting. Brains vary across individuals; tissue can deform slightly during surgery; an implant can drift as cement cures. For this reason, careful labs treat implantation as probabilistic rather than absolute: they aim carefully, they record coordinates, they test light output, and later they verify placement anatomically.

That verification matters because the implanted tip does not illuminate a neat little sphere. Tissue scattering spreads light into a gradient, and the volume that actually crosses an opsin's activation threshold depends on delivered power, wavelength, and expression level. For the animal, though, a different issue comes first: will the implant stay in place? Chronic implants ask the skull to do an unfamiliar job: act as a foundation for a rigid structure that will be tugged repeatedly. Head-cement mixtures and small screws can anchor the assembly, but nothing is indestructible. Mice scratch. Rats can be ingenious with their teeth. Social housing adds new risks: a curious cage mate can treat a ferrule like a handle.

This leads to the second, less visible choreography: training the experimental routine around the hardware. Many behavioral tasks rely on spontaneity—an animal's natural exploration, its willingness to approach or avoid, its ability to fall asleep. The tether can quietly bias all of these. If the patch cable pulls slightly upward, it can encourage a posture that makes rearing less likely. If the cable drags against the edge of a chamber, the animal may avoid that area for purely mechanical reasons, confounding spatial measures. The best behavioral optogenetics often looks, from a distance, almost ordinary: the animal moves naturally, and the hardware seems absent. Achieving that ordinariness is an engineering accomplishment.

A concrete example of how hardware and biology interlock comes from experiments that require stimulation during sleep or quiet wakefulness. The animal's head must be connected to the light delivery system without repeated handling that would itself disrupt state. The commutator must be smooth enough not to transmit tension that might wake the animal. Light power must be sufficient to recruit the targeted cells without causing heating that would alter arousal. These are not separate checkboxes. A small increase in connector friction can translate into a slight head pull; a slight head pull can change sleep posture; a changed posture can alter sleep quality; altered sleep changes the very neural dynamics under study. The clean logic of "stimulate these cells during this state" only holds if the machinery respects the animal's behavior.

Once optogenetics moved into freely moving animals, it also ran into a powerful temptation: combine control with recording. If stimulation can make a circuit do something, recording can reveal how the circuit responds and how other regions change in sympathy. But recording adds weight, bulk, and fragility. Polina Anikeeva and colleagues (2012) introduced the optetrode, a design that makes the trade-off explicit: a light guide paired with tetrodes (fine wire bundles) for extracellular electrophysiology. The light delivery element becomes a structural backbone, and the recording wires cluster around it. Done well, the arrangement allows the experimenter to perturb and listen in the same neighborhood, in the same animal, across days. Done poorly, the implant becomes heavy enough to change behavior or unstable enough to lose units.

A further step is addressability: delivering light to more than one location, or delivering distinct patterns to separate outlets, while the animal remains unrestrained. Eran Stark, Tibor Koós, and colleagues (2012) demonstrated diode probes for spatiotemporal optical control of multiple neurons in freely moving animals. The central idea is easy to appreciate: rather than a single light exit point, build an implant with multiple independently controlled outlets. The cost is complexity. Each additional channel adds cabling, connectors, potential failure points, and mechanical load. The benefit is a new class of experiments that can test interactions: stimulate one node, inhibit another, alternate patterns, or compare pathways within the same subject.

At some point, the tether itself becomes the main remaining constraint. Even with a good commutator, a cable changes the animal's life. It limits how far the animal can roam, complicates social housing, and makes certain natural behaviors hard to interpret. That pressure helped motivate wireless optogenetics: move the light source onto the animal, or at least move the last step of light generation close to the implant.

Tae-il Kim, Jordan McCall, and colleagues (2013) pushed this idea dramatically with miniature, implantable optoelectronics that enabled wireless optogenetics. The trade shifts immediately. Wireless systems can improve naturalistic movement, but they constrain power and wavelength options, and they introduce their own thermal problems: generating light in a small package inside or on top of tissue means heat must be managed with great care. Packaging must be robust against bodily fluids and mechanical stress. Power delivery, often through inductive coupling or batteries, becomes its own engineering problem. A tether supplies essentially unlimited energy and can support high-power illumination; a wireless device is freer but must be frugal.

The existence of this tether-versus-wireless trade is revealing. Optogenetics is often praised for precision, but precision is always embodied. In a freely moving animal, embodiment includes the feel of the headstage, the drag of a cable, the click of a connector, the slight torque transmitted by a commutator. For a mouse, these are not trivial. They are part of the environment, and the environment shapes behavior as surely as neurons do.

That is why seasoned optogeneticists talk about hardware as a kind of honesty test. If a manipulation produces a striking change in fear, reward seeking, or social interaction, the first skeptical question is not philosophical. It is mechanical: could the light delivery apparatus itself have nudged the animal? Was the stimulation accompanied by a subtle tug, a warming sensation, a visible flash leaking from the connector? These are mundane possibilities, but they are not far-fetched. The brain sits inside a body that is exquisitely sensitive to touch and temperature.

The best solutions tend to look boring: lighter patch cords, cleaner connectors, better strain relief, more reliable implants, careful habituation, thoughtful arena design. Boring, here, is a compliment. It means the apparatus is no longer writing its own story into the animal's behavior. The light can then do what it was meant to do: act on neurons while the animal continues to run, pause, investigate, and decide, as if it were simply living its life.

7.3 The Honest Shape of Illumination

The phrase "we stimulated the amygdala" has the satisfying clarity of a map label. It suggests a neat territory: an on-target region bright with control, surrounded by darkness. In the actual brain, light is not a paintbrush with a crisp edge. It is a spreading, fading field, shaped by tissue and hardware and the biology of expression. If optogenetics is going to support strong statements about circuits—about fear, craving, compulsion, mood—it has to be paired with a more candid picture of what illumination looks like inside living matter.

A useful way to begin is to imagine standing in a heavy fog with a flashlight. Close to the lens, the beam is bright and directional. A few steps out, it becomes a glowing haze. Farther still, it is mostly gone, not because the photons vanished, but because they have been scattered into directions that no longer carry your intent. Brain tissue behaves similarly, except the scatterers are microscopic: lipid membranes, myelin, cell bodies, blood vessels, and the fine-grained texture of the extracellular space. When light leaves the tip of an implanted guide, it starts with a geometry imposed by optics—a divergence that depends on the guide's numerical aperture (NA) and the refractive index of the surrounding tissue. Almost immediately, scattering starts to scramble that geometry. Absorption, too, begins to siphon off photons into heat and chemistry. The stimulated volume becomes a three-dimensional irradiance distribution: a cloud whose brightness decays with distance and whose boundaries are defined by thresholds rather than walls.

Those thresholds are not an aesthetic detail; they are where the experiment lives. Opsins are not triggered by a single photon arriving once. They require a certain photon flux over time to open enough channels (or pumps) to alter excitability. So the question is never simply "Where does light go?" It is "Where does light remain intense enough, long enough, to cross the activation threshold for the opsin in this particular tissue?" The answer changes with wavelength, power, pulse timing, tissue properties, and expression level, and it changes across individuals.

The brain adds a second layer of blur: opsin expression is not uniform. Even within a genetically targeted population, some neurons express more opsin than others. Some parts of a structure are infected more strongly by a viral vector, or transgene expression may vary with local cell health and developmental history. So a given irradiance field intersects a patchwork of sensitivities. Two adjacent neurons can sit in the same light and react differently. A dimmer neuron with high expression may respond where a brighter neuron with low expression does not. This is why a phrase like "we drove region X" can be misleading even when the surgical placement is perfect. The manipulation is a compound of optics and biology.

When researchers first began building behavioral optogenetics, it was tempting to treat illumination as a small, controllable sphere around the tip: choose a power, quote a radius, proceed. Over time, that confidence became harder to defend. Modeling work argued that simplified approaches could misrepresent true light spread. Some quick calculations borrowed from older, approximate frameworks (including certain Kubelka–Munk-style treatments) can make illuminated volumes look cleaner and deeper than they really are. More realistic, validated models incorporate geometry and compare predictions to measured light distributions. These details can sound like technical housekeeping until you realize what is at stake: if your effective irradiance field is smaller than assumed, you may not actually be perturbing the neurons you think you are. If it is broader, you may be perturbing more than you admit.

A striking example of how intuition can drift comes from careful estimates of scattering in cortex. A reported cortical scattering-length estimate on the order of $\sim\ 47\,\mu\text{m}$ at $473\,\text{nm}$ (blue light) implies substantially faster loss and spread of blue light than many informal assumptions used in experimental planning. In ordinary language: blue photons in cortex may lose their original directionality over tens of micrometers, not hundreds. That does not mean blue optogenetics "doesn't work." It means the light's useful structure is fragile. The illuminated region may become shallow quickly, and the notion of a tidy cone can be more psychological than physical.

Once that honest geometry is acknowledged, a set of biological complications comes into focus.

First, axons of passage. A fiber tip placed above a nucleus might illuminate not only the local cell bodies but also axons traveling through the region. Axons are cables; they can carry the consequences of optogenetic activation far from the illuminated site. If those axons express opsin—either because the opsin was expressed in the projecting neurons, or because expression included fibers passing through—light can trigger action potentials that propagate to distant targets. The outcome can be a real and meaningful circuit effect, but the spatial story becomes slippery: the illuminated location may not be the functional origin of the driven activity. This is especially tricky when interpreting pathway manipulations. A stimulation intended to bias a local microcircuit can inadvertently recruit long-range projections, creating a broader perturbation than expected.

Second, rebound and network after-effects. Inhibition is often described as simply "turning neurons off," but neurons and circuits can respond to an imposed silence with compensation. Some cells show rebound firing when inhibition ends. Networks can respond to a synchronized pause with a synchronized burst. A short inhibition epoch can therefore create a delayed excitation that is misread as a direct consequence of "release," when it is actually the circuit's own dynamics snapping back. This becomes important when stimulation is timed to behavioral events: a manipulation that ends just before an animal makes a decision can shape the decision via rebound rather than via the intended suppression.

Third, heating and other non-opsin effects. Light energy deposited in tissue does not ask whether your opsin is expressed. Time matters here in a particularly unforgiving way. Heating is a coupled spatiotemporal field; prolonged illumination or high duty cycles can raise temperature and confound neural effects with thermal ones. The same average power delivered with different pulse trains can produce different temperature dynamics. A brief, intense pulse followed by a long dark interval can let heat dissipate, while a steady high-duty cycle can accumulate warmth. The animal's brain is not a static block; blood flow carries heat away, and local vascular responses can alter that flow. The resulting temperature profile can have its own physiological consequences: ion channel kinetics shift; synaptic release changes; glial activity and vascular tone can be altered. A researcher can be confident that neurons changed their firing during illumination and still be mistaken about *why*.

A vivid vignette helps. Imagine two protocols that deliver the same total number of photons over ten seconds. Protocol A uses crisp pulses: short flashes at low duty cycle. Protocol B uses near-continuous illumination at lower intensity. The opsin might respond similarly in both cases, at least in the cells closest to the tip. Temperature may not. In Protocol A, tissue has time to cool between pulses; in Protocol B, it may warm gradually and stay warm. If an animal's behavior shifts only in Protocol B, the temptation is to attribute the effect to "more sustained neural modulation." Without thermal controls and modeling, it could just as plausibly be a small temperature rise nudging network excitability. The resulting interpretation could be elegant and wrong.

This is the point where experimental controls stop being a box to tick and become the spine of the claim. Several controls have become standard not because reviewers are picky, but because the physics and biology demand them.

A sham-light control delivers light in the same way but at a time when it should not matter for the hypothesized mechanism, or in a pattern that should not drive the opsin effectively. It tests whether the mere presence of the cable tug, the click of equipment, or the animal's perception of light leakage is contributing.

A no-opsin control uses animals that undergo the same surgery and light delivery but do not express the opsin. It probes heating, photoelectric effects at electrodes, and any general tissue response to illumination.

Whenever possible, measuring light output at the connector before and after sessions helps catch a silent culprit: gradual loss of coupling efficiency. Otherwise, a "negative result" can be nothing more than a dirty ferrule.

These practices also reflect a shift in scientific culture around optogenetics. In the early excitement, it was easy to speak as if light delivery were a simple independent variable: set power to x, illuminate for y milliseconds, conclude. As experiments became more ambitious and more consequential, illumination fields began to be treated like an experimental variable that deserves its own measurement and modeling. Open-source and replication-minded modeling efforts have grown partly out of reproducibility concerns: two labs can run "the same" stimulation protocol and still perturb different volumes because of subtle shifts in hardware, targeting, and tissue properties.

Spatial humility does not weaken optogenetics; it strengthens it. A claim such as "stimulating a small region of the basolateral amygdala triggers freezing" is less informative than "a certain pattern of activation in a genetically defined population, within a certain illumination field around the implant, biases the network toward freezing." That longer sentence is closer to reality. It admits that light is not a scalpel and that circuits are not neatly bounded organs. It also encourages better questions. If a manipulation produces a strong effect, is it driven by somata near the tip or axons passing through? Does the effect scale with power in a way that matches opsin thresholds, or does it scale in a way that suggests heating? Does the timing track expected neural dynamics, or does it look like rebound and network compensation?

Even patterned illumination does not escape these issues. Patterned beams degrade with depth; scattering blurs sculpted patterns, undermining spatial precision claims unless compensated by multiphoton or wavefront strategies, or unless illumination is delivered close enough that scatter has less room to do damage. A projected pattern that looks sharp in an optical bench setup can smear into a low-contrast glow in intact brain. The moral is not pessimism. It is that every promise of precision must be cashed in the currency of tissue optics.

There is a human reason this honesty matters. Optogenetics is often invoked in the same breath as mental illness, addiction, and the hope of targeted interventions. Those hopes are justified only if the causal stories are robust. A fragile interpretation built on an oversimplified picture of illumination can send a field chasing the wrong circuit, or the right circuit for the wrong reason. The most responsible version of optogenetics is not the one with the flashiest hardware; it is the one that treats photons, proteins, and brain tissue as partners in a single experiment and refuses to pretend that any of them behaves ideally.

The light that enters the brain does not draw borders. It makes a landscape of probability: bright near the source, rapidly fading, unevenly effective across cells, and capable of influencing biology through more routes than the opsin alone. A well-designed experiment respects that landscape and still extracts something precious from it: not a cartoon of "region X on," but a grounded statement about what kinds of neural activity, in what neighborhood of tissue, are sufficient to shift a living animal's choices.

7.4 Writing with Light, Not Just Shining It

A classic optogenetics experiment with an implanted waveguide feels, at heart, like turning on a porch light. You illuminate a small neighborhood of tissue and watch what the animal does. That approach has been extraordinarily productive, but it carries an unavoidable ambiguity: the neighborhood is not a single thing. It contains many neurons, many axons passing through, and a web of synapses whose geometry you do not control. When the animal freezes, or seeks, or forgets, it is often unclear whether the crucial cause was a particular subpopulation, a sparse set of highly connected cells, or a specific temporal pattern that the broad illumination only approximated.

The push to "write" with light begins with a simple frustration: *shining* changes many elements at once. If the goal is to understand circuits as circuits—organized patterns of connection and timing—then the manipulation needs to be organized too. Writing with light means shaping illumination in space and time so that the perturbation starts to resemble the brain's own grammar: selected cells, particular layers, defined ensembles, and stimulation patterns that can be aligned to natural activity rather than imposed as a uniform flood.

This turns optogenetics into a hybrid of genetics and optics and computation. The question is no longer merely whether you can deliver photons, but whether you can deliver them to the right addresses, at the right times, in the right combinations.

The most intuitive form of writing is to move from a single bright spot to multiple, selectively placed spots. If you can target neurons one by one, you can begin to ask a sharper kind of question: not "Is this region involved?" but "Which neurons within it carry the causal burden?" The answer matters for how we think about memory traces, fear associations, and the cellular logic of relapse. Many models of brain function assume that information is carried by sparse ensembles—small, distributed sets of neurons whose coordinated activity is more meaningful than the average activity of the region. Flooding a volume with light is a blunt way to engage such ensembles. It may recruit many neurons that are merely nearby, drowning the signal in biological noise. Writing aims for selectivity.

A major tool for this is computer-generated holography (CGH), often paired with a spatial light modulator (SLM). An SLM is a device that can impose a programmable pattern on the phase of a light wave, like a dynamic, pixelated piece of glass that can be updated electronically. CGH uses SLM-displayed phase masks to generate user-defined 2D/3D intensity patterns; it enables parallel stimulation of multiple neurons but faces efficiency and speckle constraints. "Parallel" is the key word. Instead of steering one beam sequentially, CGH can create many spots at once, allowing simultaneous activation of a chosen set of cells. In principle, it is the optical analogue of playing a chord rather than a single note.

In practice, this is where the brain's optical hostility reasserts itself. Even if you sculpt a beautiful pattern at the microscope's objective, scattering in tissue degrades it quickly in one-photon illumination. Speckle—grainy interference patterns produced when coherent light scatters—adds further irregularity. Some neurons get a little more than intended; others get less. The pattern blurs with depth, and the crisp boundaries that make the method appealing soften into something closer to a glow. This is not a failure of engineering so much as a reminder that the brain is not a clear medium.

Two-photon excitation emerged as a powerful response to this limitation. The core idea is that instead of exciting the opsin with a single photon of the "correct" color, you use two lower-energy photons arriving nearly simultaneously. The probability of this paired absorption is extremely low except at the focal spot of a tightly focused laser beam, where photon density is highest. This gives two-photon stimulation a valuable property: optical sectioning. Excitation is largely confined to a small volume near the focus, reducing out-of-plane activation. In scattering tissue, multiphoton approaches and careful optical design are used to maintain usable axial and lateral confinement.

The human appeal of this is immediate. If light can be confined to a tiny focal volume, then you can target individual neurons in three dimensions, even in relatively dense tissue. You can stimulate one cell body while leaving its neighbors mostly untouched, provided the opsin and the optical setup cooperate. The difference in experimental meaning is profound. A regional flood tests whether *some* activation within a neighborhood can drive an outcome. A cell-targeted perturbation can test whether *these particular* cells are sufficient, or whether the network requires a broader coalition.

Two-photon strategies come in several flavors, each reflecting a different compromise among speed, precision, and power. Spiral scanning traces a tight spiral over a cell body, depositing energy efficiently across the soma without lingering unnecessarily in one spot. Temporal focusing shapes a pulse so that it becomes short (and therefore intense) only at the desired plane, improving axial confinement while spreading the beam laterally. Scanless 3D-CGH combines holography and two-photon excitation, aiming to generate multiple focal spots in three dimensions simultaneously. These techniques share an uncomfortable truth: two-photon strategies trade optical power budgets against spatial precision; more targets typically require more total laser energy or more sensitive opsins. Stimulating a single neuron with careful scanning can be manageable. Stimulating fifty neurons in parallel can demand enormous peak powers, pushing against heating limits, photodamage risk, and the practical capabilities of lasers and optics.

This is the point where the physics of light meets the politics of an experiment. A researcher can specify an ideal manipulation—activate exactly this ensemble, at exactly this phase of a theta rhythm, while imaging activity—and then collide with the hard ceiling of what can be delivered safely. The brain is not a tolerant optical bench. It heats. It scatters. It bleaches fluorophores and stresses tissue. Even when a two-photon pattern is beautifully confined, the total energy required to drive many opsin molecules at depth can be substantial. The dream of writing with light therefore depends not only on better optics, but on better opsins: proteins that respond strongly to fewer photons and can be driven with wavelengths compatible with multiphoton methods.

There is a historical rhythm to this co-evolution. As soon as optogenetics made deep brain control plausible in moving mammals, new questions became psychologically irresistible. Could one tag the neurons active during a particular experience and later reactivate exactly those neurons to trigger the associated memory? Could one identify an ensemble in a fear circuit and test whether it is sufficient to evoke a defensive state? These are ensemble-level questions, and they invite ensemble-level tools. Broad illumination was enough to prove that certain pathways could push behavior. It was not enough to cleanly assign causality to a sparse pattern.

"Writing" also includes an intermediate class of techniques that do not demand a full two-photon microscope but still move beyond a single emission point. Hardware-level writing can mean dynamically addressable illumination through waveguides. A particularly elegant example uses tapered fibers: glass guides that narrow along their length so that light leaks out in a controlled way, not just at the end. By changing how light is launched into the guide—for example, which modes are excited, or which wavelengths are used—the emission zone can shift along the taper. The implant stays still;

the illuminated depth changes. This is a subtle but powerful shift in experimental design. A single implant can interrogate subregions at multiple depths without repeated surgery or multiple penetrations. It also allows a kind of "optical scanning" in depth while keeping the external hardware relatively simple compared with full holographic microscopy.

Tapered fibers also have a psychological advantage: they force clarity about geometry. A standard end-emitting guide invites a mental picture of a cone radiating from a point. A taper reminds you that emission can be distributed and controllable, and that the question "What did we stimulate?" is fundamentally about a three-dimensional irradiance distribution. It encourages experiments that map function along an axis: for instance, probing whether dorsal and ventral portions of a structure contribute differently to a learned response. These are often the distinctions that matter clinically. Anxiety, compulsion, and addiction are not properties of single dots in the brain; they are emergent properties of extended circuits with gradients of connectivity.

Writing with light becomes most conceptually thrilling when stimulation is paired with imaging in the same preparation. Imaging provides a map of activity: which neurons respond to a cue, which neurons fire during a decision, which neurons participate in a memory replay. Stimulation provides a way to test whether those observed patterns are merely correlates or are functionally important. Together they enable "all-optical" experiments: optical readout plus optical perturbation. The promise here is not only precision but feedback. If you can see the ensemble in real time, you can adjust stimulation patterns to match it, or to deliberately disrupt it, moving from generic perturbations toward targeted hypotheses about coding.

That promise is tempered by a practical tension. Imaging and stimulation often compete for spectral space and energy. Fluorescent indicators require excitation light and emit fluorescence that must be detected with high sensitivity. Optogenetic stimulation introduces intense light that can leak into detectors and overwhelm signals. Spectral separation helps, as does careful timing: briefly stimulate, then image; image, then stimulate. Two-photon methods can help by confining excitation and allowing deeper access, but they can also increase complexity and power demands. Writing with light is therefore as much about experimental choreography as the tethered setups, only now the choreography happens in microseconds and micrometers rather than in patch cords and commutators.

CGH itself brings an additional kind of honesty: it exposes how much of stimulation is computational. To generate a desired pattern, the system must calculate a phase mask. The mask is not a direct picture of the target; it is a set of instructions for how light should interfere with itself to create intensity peaks at chosen locations. If the tissue scatters and distorts the wavefront, the computed mask becomes less accurate. Correcting this requires either measuring the distortion or designing methods that are robust to it. Wavefront strategies, adaptive optics, and calibration routines are all attempts to keep the writing legible in a medium that wants to smudge it. Even when such corrections are not used, the computational framing encourages a more explicit notion of what "targeting" means: not a claim, but an engineered output whose fidelity can be tested.

The emotional core of all this technology is the desire to move from *regions* to *ensembles*. Consider an example that often comes up in conversation even outside neuroscience: addiction. People want to know why cues can feel irresistible, why relapse can occur after months or years, why craving can appear suddenly and with a bodily force. Circuit-level explanations frequently invoke ensembles in reward and stress systems—specific patterns of activity that link a cue to a motivational state. Flooding a region and seeing a change tells you that the region participates. Writing with light opens the possibility of identifying a sparse set of neurons that respond to a cue and then testing whether that set is sufficient to recreate a craving-like state, or whether disrupting that set weakens cue-driven seeking. The closer the manipulation comes to the ensemble that the brain itself uses, the less you are imposing an artificial pattern and the more you are interrogating a natural one.

Yet writing is also a warning against overconfidence. The more precise the method appears, the more seductive it becomes to claim single-cell causality. But even a single neuron is embedded in a network. Stimulating one cell can recruit its downstream partners; inhibiting it can unmask other pathways. What writing with light truly buys is not a license for simplistic conclusions, but a way to ask better-posed questions. If stimulating ensemble A produces a state and stimulating ensemble B does not, that is informative even if neither ensemble is the entire story. It narrows the hypothesis space from "somewhere in this region" to "these particular participants, with these temporal relationships."

The ultimate expression of writing would be to deliver patterns that mimic the brain's own activity patterns, not merely turn neurons on. That is technically difficult: neurons encode information in precise spike timing, bursts, synchrony, and oscillatory phase relationships. But even partial progress changes the kind of causality optogenetics can test. Instead of asking whether activity in a population is sufficient, one can ask whether a particular *pattern* is sufficient: not merely firing, but firing in a specific order, at a specific rhythm, across a specific set of cells.

This is why optogenetics increasingly resembles a dialogue with computational neuroscience. The tools are becoming capable of imposing structured perturbations, and the theories are becoming explicit about what structured patterns might mean. The brain, for its part, remains stubbornly physical: scattering blurs patterns, speckle adds randomness, power limitations impose ceilings, and tissue health imposes moral boundaries. Writing with light is therefore always a compromise between what you would like to stimulate and what you can stimulate without turning the experiment into a thermal or optical artifact.

What makes the effort worthwhile is the change in what can be said with a straight face. When illumination is sculpted to match a chosen set of neurons, and when readout confirms that those neurons responded as intended, the experiment begins to approach an older dream in neuroscience: not just correlating activity with experience, but testing whether a particular circuit element, or a particular ensemble, actually *does* the thing it seems to do. That kind of specificity is not an aesthetic preference. It is the level at which memory traces, fears, and habits are likely to be organized in the living brain: not as glowing regions, but as patterns written into networks.

Chapter 8

Seeing Cause and Effect: Recording What the Light Did

A flash of blue light may trigger a movement, a memory, or a fearful pause—but what, exactly, happened inside the brain between the flash and the behavior? The answer depends on how scientists listen back: whether they count electrical spikes, watch cells glow, compare fast optical control with slower chemical tools, and guard against artifacts that can impersonate insight.

8.1 Spikes or Glows? Two Ways of Catching a Neuron in the Act

The satisfying part of shining light into the brain is the clean sense of agency: press a button, deliver photons, and a carefully chosen group of cells changes what it is doing. The unsatisfying part is that the brain rarely announces, in plain language, what changed. To claim that light *caused* something in a circuit, you need a readout that can answer a deceptively simple question: did the targeted cells actually respond, and what did their neighbours do in reply?

Two families of tools dominate that answer. One treats the brain as an electrical organ and listens for its clicks in real time. The other treats the brain as a chemical machine and watches it glow. Both are widely used, both can be exquisitely sophisticated, and both can mislead if you forget what they are truly measuring.

Electrical recording, in its most common form, is a kind of eavesdropping. A thin electrode is placed near neurons, and it picks up tiny voltage changes in the surrounding fluid. When a nearby cell fires a spike, the electrode records a brief, sharp deflection. If you have heard a Geiger counter, you have the right intuition: spikes register like discrete events, each one occurring at a precise time. The electrode does not have to enter the cell; it is enough to be close. Because spikes are fast, the method is fast. A good recording system can tell you *when* a spike happened to within a fraction of a millisecond.

Calcium imaging speaks a different dialect. Here you typically give neurons a fluorescent molecule, often a genetically encoded calcium indicator, that changes brightness when calcium levels inside the cell rise. Calcium, for neurons, is both a messenger and a currency: it floods in through channels when the cell is active, helping trigger chemical release at synapses and switching on intracellular machinery. If calcium goes up, the indicator glows more strongly. A microscope can then record that glow from many cells at once, often laid out like a crowd in a stadium. Instead of clicks, you get pulses of brightness: slower, smoother, and spread across space.

It is tempting to translate one into the other, to treat a glow as a blurred version of a spike train, or to assume that spikes are the "real" thing and imaging is a convenient but inferior proxy. Neither is quite right. They are not interchangeable readouts; they answer slightly different questions, and they earn their authority in different ways.

Start with what an electrode gives you that a camera usually cannot:

timing. The brain cares about timing. Cells can fire together, in sequence, or in careful alternation; a few milliseconds can separate "together" from "after." When you stimulate with light, timing is often the whole point: you may be attempting to impose a rhythm on a circuit, to imitate a natural burst pattern, or to test whether one group of neurons can recruit another quickly enough to explain a reflex or a choice. If your readout cannot resolve those millisecond-scale relations, the story of cause and effect becomes smeared.

That is why electrophysiology became the early workhorse for optogenetics. When channelrhodopsin-2 began to be widely used, a natural demonstration was wonderfully direct: shine blue light, see spikes appear on cue in a recording. The trace itself looks like a mechanical proof. Even a non-expert can grasp the logic. Light arrives, the neuron answers with a spike, and the rest of the experiment can build on that handshake.

But the intimacy of an electrode comes with a price: you are listening through a keyhole. Each electrode hears only a small volume of tissue, and the voice it hears is anonymous. The electrode records voltages, not cell identities. In a typical extracellular recording, you do not automatically know whether the spikes came from an excitatory pyramidal neuron, a particular class of inhibitory interneuron, or a fibre passing by. You can infer, sometimes, from spike shape or firing patterns, but inference is not certainty. The modern response to this limitation has been engineering: high-density silicon probes with dozens or hundreds of recording sites can sample many neurons simultaneously, and clever analysis can separate mixed signals into individual "units." Yet even then, an electrode array offers a census of electrical events without a reliable name tag for each participant.

Calcium imaging flips these trade-offs. With a microscope, you can record from many cells across a field of view, and you can often know *which* cells they are. If the fluorescent indicator is expressed only in a genetically defined population, the glow already comes with a label. You can see a sparse group of cells light up in response to an event, or watch a wave of activity travel across a region, or discover that two subgroups alternate. The language is spatial and collective.

This is not merely a convenience. Many of the most urgent questions in systems neuroscience are questions about ensembles: patterns across groups of neurons that jointly represent a memory, a location, a threat, a habit. When people speak about a "circuit" in everyday terms, they often imagine a tidy chain: neuron A drives neuron B, which drives neuron C. Real circuits look more like committees and crowds. Calcium imaging is powerful because it makes crowds visible.

Yet calcium imaging is not simply "watching neurons fire." It watches calcium. Calcium rises when neurons fire spikes, but it can also rise because of subthreshold activity, synaptic input, or dendritic events that do not necessarily produce a spike at the soma. The indicator then converts calcium concentration into fluorescence in a nonlinear way, with its own kinetics: a rise time and a decay time. Even the fastest modern indicators have dynamics that are slow compared to spikes. What you see is a biochemical afterglow, an average over time of whatever the cell has been doing, biased toward recent activity.

This is why the most honest way to read a calcium trace is to treat it like a long-exposure photograph. A long exposure captures more light and reveals dim structure, but it sacrifices the crisp timing of a fast shutter. If a neuron fires a single spike, the fluorescence may rise a little and then drift back down over hundreds of milliseconds or longer. If it fires a burst of spikes, the rise is larger, but the peaks blend together. A calcium trace often tells you *how much* a cell was active over a short window, not exactly *when* each spike happened.

Scientists have become very skilled at "spike inference," using algorithms to estimate likely spike times from fluorescence. Those estimates can be quite good in some settings, especially when signals are large and noise is low. But the inference is still an interpretation layered on an interpretation: spikes influence calcium; calcium influences fluorescence; the microscope samples fluorescence at discrete times; then an algorithm guesses spikes. Each step can distort the original events. This does not make calcium imaging untrustworthy; it makes it a different measurement with a different kind of honesty.

A concrete example shows how these languages can lead to different stories. Imagine you stimulate a small group of neurons with a brief flash of light. With an electrode, you might see a cluster of spikes precisely time-locked to the flash, followed by a short silence, then a rebound. That pattern could suggest direct activation followed by inhibition, perhaps because local inhibitory interneurons were recruited. With calcium imaging, you might instead see a slow rise in fluorescence in the stimulated cells, and a delayed rise in a neighbouring population. The delay might reflect the slower kinetics of calcium signals rather than a true synaptic delay. Or it might reflect that the neighbouring population was not spiking, but receiving sustained input that raised calcium modestly. The same underlying circuit could thus look like a crisp volley-and-countervolley in spikes, or like overlapping clouds of brightness in calcium. Both can be true accounts; they simply illuminate different aspects.

Historical accidents helped cement these cultures. Electrophysiology is older, and it grew up with the idea that spikes are the fundamental events of communication. In the mid-twentieth century, the classic experiments of Vernon Mountcastle, David Hubel, Torsten Wiesel, and many others used electrodes to map how neurons respond to touch or to visual features. The craft was part surgery, part electronics, part patience: lowering an electrode by micrometres, listening for the telltale crackle that signalled a responsive cell. The reward was a kind of intimacy with time. A neuron could be heard changing its mind as a stimulus shifted by a fraction of a degree.

Imaging has a different lineage, tied to the microscope and the hope of seeing many cells at once. A key conceptual turn came from realising that if you could make activity visible, you could watch circuits operate as networks rather than as isolated points. The arrival of genetically encoded calcium indicators accelerated this vision by making it possible to label specific populations and image them repeatedly. In animals performing tasks over days, imaging allows something close to a narrative: the same neurons can be watched as an animal learns, forgets, and relearns. Electrodes can also be stable, but imaging makes stability feel natural, because the cells are literally in view.

Optogenetics, arriving into this landscape, forced a reconciliation. It demanded readouts that could keep up with intervention. If you press a light-triggered switch, you want to know whether the target neurons responded, whether they responded quickly, and whether the response propagated. A single electrode can tell you about speed and timing, but not identity and breadth. A microscope can tell you about identity and breadth, but not necessarily about speed and exact timing.

That tension shows up in a deceptively simple manoeuvre that many optogenetic studies rely on: *optical tagging*. Suppose you have expressed an opsin only in a particular genetic class of neuron. During an electrical recording, you deliver brief flashes of light and look for units that respond with short-latency spikes. Units that spike immediately are tagged as members of the opsin-expressing class. It is an elegant trick: a way of giving anonymous spikes a name.

The trick can also backfire. In a densely connected circuit, a neuron that does *not* express the opsin can still fire at short latency because it is driven by a nearby opsin-expressing neuron. If the stimulated cells are excitatory and strongly connected to local interneurons, interneurons might spike quickly in response to synaptic input, sometimes nearly as fast as the directly activated cells. Expression levels vary as well; some opsin-expressing neurons may need stronger light to fire reliably, while others respond to a whisper of illumination. The result is that "spike on cue" is compelling evidence of involvement, but not always proof of direct opsin expression. The brain, with its fast synapses and recurrent wiring, can make indirect effects look immediate.

Imaging has its own version of this problem. If you see a group of cells brighten after a light pulse, those cells may indeed have spiked. But they may also have experienced dendritic calcium events or received synaptic drive that elevates calcium without producing many spikes. Conversely, a cell could fire a few spikes with minimal calcium signal if the indicator is weakly expressed or the spikes are sparse and the imaging is noisy. Glows can be missed, and glows can exaggerate.

The essential intellectual move is to ask, each time, what sort of claim the measurement supports. An electrode supports claims about spike timing, synchrony on short time scales, and rapid causal relations. Imaging supports claims about which cells participate, how activity is distributed across space, and how population patterns evolve over hundreds of milliseconds to seconds. When light perturbs a circuit, these are distinct facets of the response.

A useful way to hold this in mind is to imagine two court reporters transcribing the same heated debate. One records every interruption and every pause, capturing the rhythm of argument but not always the identity of the speakers. The other writes down who spoke and roughly how long, capturing the social pattern of the discussion but smoothing out the exact timing. Neither transcript is "wrong." They are tailored to different questions.

The practical consequences show up in experiment design. If the hypothesis is that a particular group of neurons drives a downstream target with a precise delay that matches a behavioural reaction time, you will gravitate toward spikes. If the hypothesis is that a memory is represented by a sparse ensemble distributed across a region, and you want to know how that ensemble changes with learning, you will gravitate toward imaging. If the hypothesis involves both timing and identity, you either combine methods or accept that some part of the story will be indirect.

It is worth pausing on why "indirect" is not an insult here. Many of the most trusted measurements in biology are indirect, from blood oxygenation as a proxy for metabolic demand to hormone levels as a proxy for internal state. The key is to know the transfer function: how the measured signal relates to the underlying event. For calcium imaging, the transfer function includes indicator kinetics, calcium buffering, and microscope sampling rate. For extracellular spikes, the transfer function includes the geometry of the electrode relative to the cell, the filtering of the amplifier, and the algorithms used to separate units. Both have their distortions; both require calibration and humility.

A mid-way snapshot of the two approaches helps keep their trade-offs straight:

Once you internalize that these are two languages, a host of recurring confusions become easier to avoid. Consider the common desire to know whether optogenetic stimulation "increased activity" or "decreased activity." With spikes, the phrase is usually literal: firing rate went up or down, spikes became more or less likely at particular times, bursts became more or less common. With calcium, an "increase" might reflect more spikes, but it might also reflect longer depolarizations, more synaptic input, or changes in calcium handling inside the cell. In some experiments, calcium signals can even saturate: when a neuron is very active, additional spikes do not make it much brighter. What looks like a plateau in fluorescence might conceal a frenzy of spikes, just as a microphone can clip when a concert becomes too loud.

Now add light to the mix. Optogenetic stimulation can itself interact with the readout. With electrophysiology, illumination can induce photoelectric artifacts at electrodes: brief voltage transients caused by light interacting with metal and tissue interfaces. These are not spikes, but they can masquerade as sharp events if you are careless. With imaging, the stimulation light can overlap spectrally with the indicator, causing cross-talk: the very photons meant to drive an opsin can excite fluorescence or bleach the indicator. Engineers and experimentalists spend enormous effort choosing wavelengths, timing pulses, and designing optical paths to keep actuation and measurement from interfering with each other. The need for that care is a reminder that, in these experiments, light is both the question and a potential source of false answers.

Anecdotes from early optogenetic work often have a similar arc: a lab successfully expresses an opsin, delivers light, and sees an effect. Then comes the harder part, the part that separates a dramatic demonstration from a stable scientific claim: figuring out *what changed where*. A group might begin with electrode recordings because they are immediate and convincing, then turn to imaging to understand which cells are coordinating. Or a lab might start with imaging to find candidate ensembles and then add electrophysiology to test whether the timing supports a particular circuit mechanism. The field's maturity can be measured by how often papers use both languages, and by how explicitly they translate between them.

Translation, though, is never perfect. If you are watching calcium signals from dozens of neurons during a decision, you might see a handful brighten before the action and conclude they are driving the choice. An electrode might reveal that these cells do not spike earlier; instead they receive earlier input that raises calcium but only later turns into spikes. Or you might see the opposite: sharp spiking changes with little calcium signal in some cells, perhaps because their calcium dynamics are unusual or because the indicator expression is weak. Each method can correct the other's blind spots, but each can also tempt you into overconfidence if you treat its output as a complete portrait.

Perhaps the most human way to summarize the trade-off is to ask what you would want if the brain were your patient. If someone's memory fails, or fear becomes overwhelming, or addiction hijacks choice, you might care about which circuits are involved: which populations are active, how widespread the disturbance is, whether the same cells participate day after day. Imaging excels at that sort of map. But if you were trying to understand a sudden panic attack, a flashback, or the split-second loss of control that makes a relapse possible, you might care about timing: which events come first, what pattern triggers what cascade. Spikes excel at that sort of clock.

In the end, neither spikes nor glows are the brain's "true" voice. They are translations, each faithful to a different aspect of the original. Optogenetics provides the push; these readouts provide the reply. If you listen only for clicks, you can miss the choir. If you watch only for glow, you can miss the beat.

8.2 The Brain Answers Back

Shining light into the brain can feel like the cleanest experiment imaginable. You choose a genetically defined population, you deliver a pulse, and you watch an animal freeze, turn, lick, or stop. The temptation is to treat that visible action as the brain's answer.

But behaviour is a loud, slow output from a system that talks to itself in fast, layered ways. A rat that freezes after a pulse of light might have frozen because the targeted cells fired, because they silenced a neighbour, because they triggered a downstream region two synapses away, or because the stimulation disrupted an ongoing pattern the animal needed to keep moving. Without listening in on the circuit, it is surprisingly easy to tell a causal story that is tidy, emotionally satisfying, and wrong.

The key shift in modern causal neuroscience is to treat stimulation and measurement as a single coupled act. Light is not the conclusion; it is a question posed in photons. The reply arrives as patterns elsewhere: spikes that appear or disappear, oscillations that tighten or loosen, whole regions that come online together or fall out of step. Those replies then reshape the explanation you are willing to accept. When that loop is tight, optogenetics becomes less like a magic trick and more like a disciplined interrogation.

A useful place to begin is the most basic form of "hearing the answer": recording from the very cells you are stimulating. This sounds almost too obvious, yet it is the first safeguard against self-deception. Opsin expression varies from cell to cell, light spreads unevenly through tissue, and neurons can be in states where they are easy or hard to excite. If you do not check whether the targeted cells actually changed their firing, a behavioural result can float free of its supposed mechanism.

In the earliest optogenetic demonstrations, that check often took the form of an "optrode": an electrode paired with an optical fibre so the experimenter could deliver light and record spikes at the same site. The logic is intimate. A pulse of light is delivered. Spikes appear with a short latency. The circuit has, at least locally, done what the design intended. Yet even this intimacy is not a full explanation, because local circuits are social. A neuron that fires can recruit inhibition, drive recurrent excitation, or release neuromodulators that reshape excitability in the neighbourhood. The direct effect of light can be rapidly wrapped in a network effect, so that a tool designed to excite can paradoxically reduce firing elsewhere, or even reduce firing in the very population you thought you were driving once feedback inhibition ramps up.

This is where pairing optogenetic perturbations with high-density recording became transformative. Instead of eavesdropping on a handful of units, silicon probes can sample many neurons across layers or across a small region, letting you see how the imposed pattern is transformed by local connectivity. Work associated with György Buzsáki and Eran Stark (2015) helped crystallize this approach: stimulation becomes meaningful when you can watch the immediate physiological echo across a microcircuit. You learn which responses are locked tightly to the light pulse, which ones appear with synaptic delays, which cells are suppressed, and how quickly the circuit returns to baseline.

That last point is easy to underestimate. Circuits have inertia. Some rebound quickly; others carry the perturbation forward as a change in state. Two protocols can deliver the same number of spikes and still have radically different consequences if one aligns with a local rhythm and the other scrambles it. Recording is how you discover whether your intervention acted like a tap on the shoulder or like a shove that changed the room's mood.

Once you accept that the brain's reply can be distributed, the most interesting answers often arrive from places you did not illuminate. A classic illustration of this scaling up came with optogenetic fMRI, often abbreviated ofMRI. In a widely cited demonstration, Jin Hyung

Lee and collaborators (2010) stimulated genetically defined neurons and measured BOLD signals across the brain. BOLD, the blood-oxygen-level-dependent signal used in fMRI, is not a direct readout of spikes; it is a haemodynamic proxy that reflects local changes in oxygenation and blood flow driven by neural and metabolic activity. It is, in other words, a reply delivered through the vascular system as much as through synapses.

That sounds like a step away from causality, yet it answered a crucial question that local electrodes cannot: where does the perturbation travel? Stimulating a cell type in one region can light up a connected set of downstream nodes, revealing functional pathways that match anatomy in some places and surprise it in others. ofMRI also made a subtle point tangible to non-specialists: the causal footprint of optogenetics need not be confined to the illuminated volume. A brief optical perturbation can reorganize a network.

At the same time, ofMRI is an object lesson in layers of inference. If a BOLD signal rises in a distant area after stimulation, it suggests that something in that distant area changed, but it does not specify what. Increased BOLD can accompany increased spiking, but it can also reflect heightened synaptic input, local inhibitory activity, or shifts in glial and vascular signalling. So ofMRI is most persuasive when it is anchored to faster recordings that verify local spiking and when its spatial patterns are interpreted with the humility appropriate to a proxy.

This is the broader pattern: recording turns the brain's answer into constraints on interpretation. A perturbation that is followed by a rapid spike response in a downstream target supports a story about fast synaptic transmission. A perturbation followed by a slow, widespread BOLD response suggests a more global reconfiguration, perhaps involving recurrent loops and neuromodulatory tone. The experimentalist's job is not to crown one measurement as king, but to assemble a chain of replies that makes alternative explanations increasingly implausible.

Causal claims become sharper when they are phrased as tests rather than as demonstrations. Two words do much of the philosophical heavy lifting here: *sufficiency* and *necessity*.

Sufficiency asks: if you artificially drive a pathway, can it produce an outcome? If activating a projection from region A to region B can trigger freezing, then that projection is capable of producing freezing. That is a powerful fact, especially when it identifies a circuit that is otherwise buried inside the brain.

Necessity asks a tougher question: does the brain normally rely on that pathway to produce the outcome? If you silence that same projection during a natural threat, does freezing diminish? A pathway can be sufficient without being necessary, much as a spare key can open your door without being the key you usually use. Optogenetics made sufficiency experiments easy and dramatic, and for a while the field was intoxicated by them. Recording, and the discipline it encourages, pulls you toward necessity and toward mechanistic explanations that can survive more hostile tests.

The interplay between sufficiency and necessity is not just academic; it is where many debates about psychiatric relevance live. Consider a circuit that, when stimulated, produces an anxiety-like avoidance pattern. That tells you the circuit can push the system toward avoidance. It does not yet tell you that the circuit is hyperactive in anxiety disorders, or that it drives avoidance in everyday anxious states. To reach toward those claims, you need to watch the circuit during naturalistic behaviour, quantify its patterns, and then perturb it in a way that is timed to those patterns. The brain's reply to the perturbation becomes evidence not only that the circuit *can* do something, but that it *does* do it under the conditions of interest.

This is where closed-loop experiments enter, and they are among the most conceptually revealing uses of optogenetics. In a closed-loop setup, the experiment does not simply deliver a pre-programmed train of light pulses. It measures a signal from the brain, detects a particular pattern, and triggers stimulation contingent on that pattern. The loop can be fast or slow, but the principle is the same: the brain is not treated as a passive recipient of light; it is treated as a dynamical system whose current state determines how it will respond.

A striking application is seizure interruption. Epileptic seizures can emerge as runaway synchrony, a pathological rhythm that recruits large populations. In some experimental models, electrodes detect the onset of a seizure-like pattern and trigger optogenetic inhibition of a relevant population, attempting to abort the event before it generalizes. When such a system works, it is not merely a technological feat. It is a strong causal statement: the detected pattern is not just correlated with the seizure, it is part of the process that sustains it, and intervening at the right moment can change the trajectory.

When such a system fails, it is equally informative. Sometimes it fails because the biomarker is not reliable: the detector triggers too late, or it confuses harmless rhythms with dangerous ones. Sometimes it fails because of latency: by the time the system decides and the light arrives, the network has already tipped into a new state. And sometimes it fails because recording is contaminated by the stimulation itself. Light can induce electrical artifacts at the recording site, creating a feedback catastrophe where the act of stimulation masquerades as the signal that calls for more stimulation. The loop becomes a hall of mirrors.

These failure modes are not embarrassments to be hidden; they are reminders of what causal control truly demands. The tighter the loop, the more it forces every link in the chain to be honest: detection must be grounded, stimulation must be specific, artifacts must be suppressed. Research and development on transparent or low-artifact electrode designs is not glamorous, but it is foundational for closed-loop claims, because without clean recordings you cannot trust the brain's reply. Engineering, here, is epistemology.

Closed-loop logic also clarifies a subtler point about optogenetic causality: a circuit's response is often state-dependent. The same pulse of light can have very different effects depending on whether the animal is asleep or awake, anxious or calm, exploring or freezing, focused or distracted. Circuits are not just wires; they are living tissues with ongoing activity patterns and neuromodulatory context. Recording reveals that context. It tells you whether your stimulation arrived during a local oscillation that gates synaptic transmission, or during a moment of spontaneous population silence, or while the target cells were already firing near their ceiling. In other words, it tells you whether you asked your question at a moment when the circuit was able to answer clearly.

Even in open-loop experiments, where stimulation is delivered on a schedule, this state dependence can turn tidy interpretations into traps. Suppose light drives a burst of spikes in a targeted population, and behaviour changes. It is tempting to say the burst *caused* the behaviour. But perhaps the burst also reset an ongoing pattern, or recruited inhibition that suppressed a competing pathway, or triggered a downstream neuromodulatory nucleus that shifted arousal. With simultaneous recording, you can look for signatures of these possibilities: a telltale suppression of neighbouring units, a delayed recruitment of a downstream area, or a change in rhythmic coordination. You can then adjust the perturbation to test competing explanations: change the timing, change the pattern, stimulate the projection rather than the soma, or stimulate during one brain state but not another. The explanation becomes something you refine against the brain's replies, not something you assert because a button was pressed.

This refinement is where optogenetics begins to resemble older traditions of physiology, but with sharper tools. In classical physiology, one stimulates a nerve and measures a muscle twitch, then alters conditions and observes how the relationship changes. Optogenetics extends the same logic inward: stimulate one cell type and record the population response, then alter timing and observe how the response reorganizes. The brain is answering back not only with what it does, but with how its answers change when you ask the question in a new way.

The most productive mindset is to treat every perturbation as a probe with a known shape and a limited reach. Light can be delivered with exquisite timing, but the circuit's response is shaped by connectivity, thresholds, and ongoing patterns. Recording is what makes those shaping forces visible. It is also what keeps causal language honest. A statement like "this pathway drives fear" becomes more than a headline when it is tied to a measurable cascade: the targeted neurons fire, a downstream population is recruited at a particular delay, competing populations are suppressed, and a network signature associated with threat is altered. Each element is a reply from the system, and together they form a story that can withstand scepticism.

If all this sounds demanding, it is because it is. The reward is that the brain starts to look less like a black box and more like a machine whose internal logic can be tested. Not perfectly read, not fully controlled, but engaged in a dialogue. You shine light, the circuit answers in spikes and patterns, and the quality of the answer determines whether the explanation is a slogan or a mechanism.

8.3 Chemogenetics in Slow Motion

Light has a certain rhetorical force in neuroscience: it is crisp, immediate, and easy to picture. A flash arrives; a neuron responds. Yet many of the brain states people most desperately want to understand are not flashes. Craving can swell for minutes before it tips into action. Stress can simmer for hours, colouring perception and memory. Mood disorders can feel like the weather inside the mind: not a single event, but a sustained condition that changes what everything else means.

A tool built for speed can still be used to study slow states, but it is not always the most natural fit. Sometimes what you want is not a precise timestamp for each spike, but a gentle, prolonged push that biases a circuit for long enough to see what kind of animal emerges on the other side. Chemogenetics was designed for that kind of question.

The best-known chemogenetic approach is based on DREADDs, short for *Designer Receptors Exclusively Activated by Designer Drugs*. The idea is almost mischievously simple. Instead of adding a light-sensitive protein to a neuron, you add an engineered receptor—a molecule that sits in the cell membrane and changes the cell's excitability when it binds a specific chemical. The chemical is chosen, or designed, to be inert (or nearly inert) in the brain unless that engineered receptor is present. Give the chemical, and only the cells carrying the receptor should respond.

Where optogenetics speaks in flashes, DREADDs speak in doses. You inject (or sometimes give orally) a ligand, the receptor's binding partner, and wait as it is absorbed, crosses into the brain, binds receptors, and triggers intracellular signaling. The onset and offset are therefore constrained by pharmacology and downstream signaling cascades. Chemogenetics (DREADDs) offers remote modulation without implanted light hardware but is constrained by ligand pharmacology and downstream signaling, typically operating on slower onset/offset timescales than optogenetics. That slower pace is not a flaw; it is a feature when the hypothesis itself is slow.

This changes the feel of an experiment. With optogenetics, a mouse might receive a train of pulses at a particular moment in a task, and you can ask whether a circuit's contribution depends on timing. With chemogenetics, you might administer a ligand before a behavioural session and ask whether the animal's overall policy shifts: does it explore less, persist longer, generalize fear more broadly, relapse more readily? You are perturbing a background condition rather than inserting a precisely timed event.

It is hard to appreciate how liberating this can be until you remember the physical burden of light. Optical experiments often require implanted fibres, head-mounted hardware, tethers, and careful alignment. Those complications are worth it when timing and spatial precision are central to the question. But when the scientific aim is to bias a widely distributed population across a large volume—say, a neuromodulatory system that sends diffuse projections—a chemical actuator can be less intrusive and sometimes more faithful to biology.

Chemogenetics also fits naturally with a certain kind of clinical imagination. Drugs are already how medicine reaches the brain. A technology that modifies the brain's response to a drug, rather than requiring an implanted light source, feels closer to therapeutic translation, at least conceptually. Even if DREADDs themselves are not destined to be used routinely in humans, the logic they embody—remote control of defined cell types via pharmacology—has influenced how people think about future interventions.

Yet chemogenetics comes with its own traps, and they are not subtle. Optogenetics forces you to confront optics and hardware. Chemogenetics forces you to confront the messy realities of drugs: absorption, distribution, metabolism, off-target binding, and variability across individuals.

A major controversy: systemic CNO can be converted to clozapine and/or may have limited brain penetration; converted clozapine can drive DREADD activation and also has endogenous targets, requiring careful controls. This single sentence contains a whole education in why causal inference is hard.

For years, a commonly used DREADD ligand was clozapine-N-oxide, usually called CNO. It was attractive because it was widely described as pharmacologically inert in animals that do not express DREADDs. In the idealized picture, CNO was a key that fit only the engineered lock. But biology does not always honour ideals. Evidence accumulated that after systemic injection, CNO can be converted in the body into clozapine, a drug with many endogenous targets in the brain. Clozapine, famously used as an antipsychotic, is not a clean tool. It binds promiscuously and can alter behaviour even in animals without DREADDs.

This creates a nightmare scenario for interpretation: you administer CNO to animals expressing DREADDs, observe a behavioural change, and attribute it to the engineered receptors. But the behavioural change might partly reflect clozapine's native pharmacology. Worse, if clozapine is the true active agent for DREADD activation in vivo, then dose becomes a double-edged sword: higher CNO doses might produce stronger DREADD effects, but also stronger off-target effects via clozapine. The same behaviour could be shaped by two overlapping mechanisms, one intended and one accidental.

The field's response has been instructive, because it reveals what "controls" really mean when the tool is a drug. If a mouse changes its behaviour after ligand injection, the first question is no longer simply "did the receptor work?" but "what else did the ligand do?" Field guidance stresses DREADD-free ligand controls and acknowledges that off-target effects can be context-dependent; debates persist over when/where CNO/clozapine effects matter behaviorally. A clean design often needs multiple comparison groups: animals with

DREADDs receiving ligand, animals without DREADDs receiving ligand, and ideally animals with DREADDs receiving a vehicle injection. Even that may not be enough if stress, arousal, or task demands interact with the drug's off-target profile.

A vivid way to see the problem is to imagine a study of craving. Suppose a lab expresses an inhibitory DREADD in a population thought to promote drug seeking. On test day, they inject CNO, place the animal in a context associated with drug availability, and observe reduced seeking. It is a compelling narrative: silencing those neurons reduces craving-like behaviour. But if CNO is converted to clozapine, and clozapine itself reduces locomotion or changes motivation, then the decrease might reflect sedation or altered reward processing unrelated to the targeted cells. The animal may appear "less craving" simply because its whole brain has been chemically nudged.

This does not mean the experiment is doomed. It means the burden shifts: you must show that the ligand has minimal effects in animals lacking DREADDs, at the chosen dose and in the chosen task. Even then, comparative PK/PD work shows that ligand metabolism/distribution can vary with biological variables (species/strain/sex/health), complicating dose selection and reproducibility across labs. A dose that behaves acceptably in one mouse strain might produce noticeable off-target effects in another. A protocol that seems clean in young animals might behave differently in older ones or in models of disease. Chemogenetics can feel like a purely genetic trick, but it is pharmacology all the way down.

These issues have driven the development of alternative ligands and a gradual professionalization of chemogenetic practice. DCZ is introduced as a potent, selective, and faster-acting DREADD agonist with utility in mice and nonhuman primates, addressing some legacy ligand limitations. "Faster-acting" here is a relative term: it still does not compete with optogenetics' rapid pulses, but it can narrow the gap enough to support experiments where a tens-of-minutes delay would be intolerable. Its reported selectivity and potency also mean that lower doses may achieve robust receptor occupancy, potentially reducing off-target binding.

Even improved ligands do not abolish the need for discipline. Alternative ligands such as C21 can show dose-dependent tradeoffs between DREADD selectivity and off-target outcomes, making validation and titration essential. In practice, this means something slightly unromantic but crucial: dose-response curves, behavioural assays in DREADD-free animals, and sometimes direct measurement of ligand and metabolite levels. Chemogenetics invites the same kind of care that pharmacologists take for granted, but which neuroscientists trained on optics sometimes learn only after being surprised.

The payoff for this caution is a tool that can ask questions optogenetics struggles to ask cleanly. Consider a circuit hypothesized to regulate a persistent internal state like vigilance. With light, you might be able to push the circuit on and off, but maintaining a stable optical modulation for an hour can introduce heating, phototoxicity, and behavioural disruption from tethers and hardware. With DREADDs, you can modulate the population for the duration of a behavioural session with no fibre, no headstage, and no requirement that the animal tolerate repeated illumination. That can make experiments feel more ethological: the animal is moving through a more naturalistic world, while its circuit is quietly biased.

Chemogenetics also fits well with questions about *baseline* rather than *trigger*. Optogenetics is superb for identifying triggers: the precise pattern that tips a decision, the pulse that can substitute for a cue, the spike timing that recruits a downstream target. Chemogenetics is better for probing baseline settings: what happens when a circuit runs a little hotter or cooler for long enough that the rest of the brain adapts? Addiction and mood are full of such baseline shifts. A person with depression may not be unable to feel pleasure in a single moment; they may be living with a chronic reweighting of what feels worth doing. A person in recovery may not relapse because of one cue alone; they may relapse because the background balance between craving, restraint, and stress has drifted for weeks. Slow tools, used wisely, can get closer to those kinds of mechanisms.

There is, however, a conceptual price to slow control: it becomes harder to separate primary from secondary effects. When you modulate a population for minutes or hours, the rest of the brain has time to compensate. Synapses can change release probability; inhibitory neurons can adjust their gain; neuromodulators can shift. The observed behaviour might be the result of the direct effect on the targeted cells plus the network's attempt to restore equilibrium. That does not invalidate the result, but it changes what kind of causal claim it supports. It is often a claim about *state* rather than *moment*: biasing this population shifts the organism toward one mode of operation.

This is where the earlier logic of "the brain answers back" becomes especially valuable. Even when using chemogenetics, recording can prevent overconfident storytelling. If you administer a ligand and then record spikes or calcium signals from the targeted region, you can confirm that the intended modulation occurred, estimate its time course, and see how other local populations responded. Without that, the drug becomes a black box: you assume it acted, you assume it acted selectively, and you assume its time course matched your intuition. The brain may politely decline all three assumptions.

A historical echo sits behind this. For decades before optogenetics, neuroscientists relied on electrical stimulation and on lesions. Electrical stimulation was fast but imprecise; lesions were slow and irreversible. Chemogenetics can be seen as a modern cousin of those older methods: less temporally sharp than light, but far more specific than a lesion, and reversible without hardware. It inherits the strengths and weaknesses of interventions that act through biology's own chemistry. It also inherits the need for humility about what a systemic manipulation can do.

When chemogenetics is used well, it can feel like adjusting the thermostat rather than flicking a switch. Turn down a population that amplifies stress responses, and an animal may become more willing to explore uncertain environments. Turn up a population that supports sustained attention, and the animal may persist longer on a difficult discrimination. These are not theatrical, instantaneous transformations. They are shifts in probability, in mood, in patience. They resemble the way many psychiatric drugs work when they work at all: not as mind-control rays, but as biases that make certain states easier to enter or harder to sustain.

The slow pace also invites a kind of experimental honesty. If a behavioural change appears only after twenty minutes, that is not a failure; it is part of the signal. It suggests the relevant mechanism is not a single spike volley but a cascade of signaling and network adjustment. If the effect lasts for hours, that too carries information: perhaps the targeted cells sit in a position to stabilize a state, or perhaps the downstream network takes time to relax back. In chemogenetics, timing is still informative; it is simply timing on a human scale rather than a spike's scale.

The deepest lesson from comparing light and ligand is that speed is not merely a technical brag. It is a way of choosing what kind of causality you are testing. Optogenetics asks, in effect, "can this circuit move the brain right now, on the timescale of decisions and reflexes?" Chemogenetics asks, "what kind of brain do you get when this circuit has been nudged for long enough to change the background?" Both questions are legitimate. Both can illuminate memory, fear, and addiction. And both can mislead if the intervention is treated as self-explanatory rather than as a prompt that must be answered, measured, and challenged.

A final, practical thought brings the slow tool back to the human stakes. People living with mental illness rarely experience their symptoms as single instants. They experience them as days that drag, nights that do not settle, urges that return, fears that refuse to extinguish. A technology that can bias a circuit gently over time, without the drama of wires and flashes, is not just a convenience for the laboratory. It is a reminder that causality in the brain often unfolds at the pace of a life being lived.

8.4 When Light Lies

The seduction of optogenetics is its apparent cleanliness. A pulse of light is delivered at a chosen moment, aimed at a chosen place, in an animal whose genes have been arranged so that only certain neurons should care. If the animal's behaviour changes, the mind reaches for a satisfying sentence: *those neurons caused that action.*

The trouble is that light is not only a messenger. It is also energy, heat, and sometimes an irritant. Hardware is not only a conduit. It is also weight, stiffness, and a source of electrical mischief. Viruses and engineered proteins are not only labels. They can be biological interventions in their own right. Even a perfectly targeted perturbation can be misread if the animal's behaviour is altered by the setup rather than by the brain.

Rigor, in this setting, is not bureaucratic box-ticking. It is the price of admission for strong causal language. The more convincingly an experiment seems to show cause and effect, the more urgently one must ask whether a simpler, uglier explanation could produce the same pattern.

The four hazards in that map are not exotic edge cases. They are common enough that experienced labs often talk about them the way a sailor talks about reefs: not as abstract possibilities, but as places you learn to steer around.

Begin with heat, because it is the most literal way light can lie. Photons absorbed by tissue do not vanish; some of their energy becomes heat. Illumination can heat tissue; models quantify how heating depends on power, duration, wavelength, and geometry, and why prolonged/high-duty-cycle protocols are highest risk. That last phrase contains a practical warning. Short, sparse pulses at modest power often add negligible heat. But an experiment that moves from "a few brief flashes" to "sustained trains for minutes" can cross a threshold where heating becomes biologically meaningful.

A small temperature change in the brain is not innocuous. Ion channels open and close at rates that depend on temperature; synapses release neurotransmitter with probabilities that can shift; blood vessels dilate; glial cells respond. If you heat a local region, you may change the excitability of many cells, including those that do not express an opsin. In the worst case, you can produce a behavioural or physiological change that looks like a circuit manipulation but is, in part, a thermal perturbation.

This is why community-facing summaries emphasize that temperature modeling and validation should be part of experimental design, not an afterthought. Modeling here is not an academic flourish. It is a way to anticipate whether your chosen power and pulse pattern plausibly warms tissue by a trivial amount or by a level that could confound interpretation. Validation can be as simple as demonstrating that opsin-negative animals receiving the same illumination do not show the effect you are claiming, or as direct as measuring temperature changes under conditions close to the experiment.

Thermal confounds can be emotionally disappointing because they feel like a betrayal: you wanted to manipulate a circuit, not run a tiny heater. Yet the point is not to banish heat from the story, but to be honest about when it could be the story. If a behavioural effect persists in opsin-negative animals, the most elegant neuronal narrative in the world will not rescue the claim.

The second hazard is less intuitive and, for that reason, even more treacherous: photoelectric artifacts. Light can induce photoelectric artifacts on electrodes and bioelectronic interfaces; artifact amplitude and shape depend on material and optical parameters, and mitigation requires informed design choices. If you record spikes while stimulating with light, a flash can produce a stereotyped voltage deflection that is not a spike but can look like one if you are eager or inexperienced. The artifact can occur with precisely the timing you would expect from direct activation: right after the light pulse, reliably, repeatably. It can therefore be mistaken for the cleanest possible evidence that a neuron fired.

This is not a purely theoretical worry. It is one reason "simultaneous stimulation and recording" is, in practice, a careful engineering and analysis problem rather than a simple combination of two devices. Engineering solutions (e.g., transparent electrodes) target these artifacts to enable reliable simultaneous stimulation and recording; without mitigation, artifacts can be mistaken for neural events. Sometimes the solution is physical: changing electrode materials, adding shielding, altering the geometry so light does not strike the recording surface. Sometimes it is optical: changing wavelengths, reducing intensity, shaping light delivery. Sometimes it is analytical: recognizing the artifact's signature across channels and rejecting it. Often it is a combination.

A sobering thought experiment shows why this matters for claims about causality. Imagine a closed-loop system meant to detect a pathological pattern and deliver light to interrupt it. If the detection electrode picks up light artifacts, the system can begin to "detect" its own stimulation, delivering more light, producing more artifacts, and spiraling into a self-fulfilling loop. In that scenario, the technology is not merely wrong; it can become actively destabilizing. The moral is that clean signals are not a luxury. They are a safety requirement whenever stimulation depends on measurement.

The third hazard sits inside the biology: expression effects. To make neurons light-sensitive, researchers commonly use viral vectors to deliver opsin genes, or they rely on transgenic strategies. Either way, the neuron is being asked to manufacture a foreign protein, insert it into its membrane, and sometimes express it at levels far beyond what it would normally devote to any single ion channel. In many cases neurons tolerate this well, and the success of the field testifies to that. Yet it is not guaranteed. Overexpression can stress cells, alter membrane properties, or change how they respond to inputs. Viral infection can trigger immune responses or inflammation. Even if these changes are subtle, they can matter if your behavioural effect is also subtle.

Expression effects have a special capacity to mislead because they can masquerade as circuit specificity. If an opsin-expressing group behaves differently from a non-expressing group, you might be tempted to attribute the difference to the optogenetic manipulation. But perhaps the expressing group's baseline physiology has shifted because of the vector, the promoter used, or the metabolic burden of expression. The remedy is not paranoia; it is matched controls. A "viral control" that expresses a fluorescent protein but not an opsin can help isolate the effect of expression and surgery from the effect of light-driven currents. Titrating expression and validating cell health with physiological measures makes the causal story sturdier.

A related interpretive confound arises even when expression is healthy: network recruitment. Optical tagging and circuit interpretations can be confounded by fast synaptic recruitment and heterogeneous opsin expression; a light-evoked spike is not always evidence of direct opsin expression in the recorded unit. This is a subtle point with big consequences. If you shine light and a nearby neuron spikes quickly, the simplest conclusion is that it expressed the opsin. But neurons talk to each other fast. An opsin-expressing excitatory cell can recruit an interneuron with a short delay; an interneuron can then suppress others. In dense circuits, the first spikes you record after light could be a mixture of direct and indirect effects. Without careful latency analysis, pharmacological checks, or complementary labeling, a unit can be misidentified. Then the entire downstream interpretation—which cell type did what, and when—tilts off its foundation.

The fourth hazard is the most mundane and, in animal behaviour, often the most powerful: behavioural disruption from the apparatus and from light as a sensory event. An implanted fibre adds weight and stiffness. A tether can tug. A headstage changes how an animal navigates tight spaces. These are not side issues if your behavioural readout depends on fine movements, exploration, or social interaction. An animal that moves less after "stimulation" might simply be moving less because the tether is annoying, because it learned that the light predicts something, or because a subtle heating sensation is aversive.

Light itself can also become a cue. Even if the illumination is delivered deep in the brain, some light can leak or scatter; the animal can sometimes see it, or perceive it indirectly. A change in behaviour might then be a response to a cue rather than a response to a circuit perturbation. This is why wavelength controls and sham stimulation are not fussy extras. If the same behavioural change appears when you deliver light at a wavelength that does not activate the opsin, the story changes dramatically.

There is an additional twist that feels almost unfair: blue light can have biological effects even without opsins. Blue light can produce biological effects even without opsins (e.g., activity-regulated gene expression changes in vitro), supporting strict light-only controls and careful choice of wavelengths/intensities. The exact mechanisms can vary with preparation, intensity, and duration, but the existence of such effects forces a disciplined habit: never assume that "light is inert" in the absence of an opsin. Light is a physical intervention. If the protocol uses high power, long durations, or repeated sessions, the need for opsin-negative, light-delivered controls becomes non-negotiable.

A historical lesson sits quietly behind these modern concerns. When electrical stimulation was first used to map brain function, it too seemed clean: apply current, observe movement or sensation. It took time to appreciate how often the current spread, how often it activated fibres of passage rather than local cell bodies, how easily it could create unnatural synchrony. Optogenetics solved some of those problems, but it did not abolish the general principle: interventions create side effects, and side effects often mimic mechanisms.

A more contemporary anecdote is familiar to many trainees. A student records from a brain region while delivering light pulses and sees beautiful, time-locked "responses" on every trial. The traces are so consistent that the student feels an almost cinematic thrill: the brain is obeying. Then a control is run with the electrode slightly repositioned, or with the fibre angled, or in an opsin-negative animal, and the "responses" persist with the same shape. The supposed spikes were an optical artifact. The thrill collapses into embarrassment, followed by something more valuable: a permanent skepticism that improves every later experiment.

This is what rigor buys. It buys the right to be excited when a result survives attempts to break it.

Controls in optogenetics work best when they are tailored to a specific worry rather than performed as ritual. If heating is the concern, the control should match illumination power and duty cycle in opsin-negative tissue, and the protocol should be evaluated with thermal modeling or measurement. If photoelectric artifacts are the concern, the control should record with the same electrodes and illumination but in conditions where no spikes should occur, or with light delivered to an inert region, so that artifacts can be catalogued and mitigated. If expression effects are the concern, the control should use matched viral delivery and expression without opsin function. If behavioural disruption is the concern, the control should match the mechanical burden: mock implants, dummy tethers, and counterbalanced handling.

Each of these controls is, in its own way, an attempt to separate three overlapping events: the intended optical manipulation of opsins, the physical consequences of illumination and hardware, and the animal's interpretation of the situation. Optogenetics is powerful because it can be targeted, but targeting does not guarantee truth. Truth emerges when the same conclusion holds after the most plausible alternative explanations have been given their best shot.

There is also a deeper, more philosophical reason to take these confounds seriously: brains are meaning-making machines. An animal in an experiment is not a passive substrate. It learns that certain contexts predict odd sensations, it adjusts its strategy, it anticipates, it becomes cautious or bold. A tether can become part of the environment the animal navigates; a repeated light pulse can become a predictor. Without careful design, the experiment can drift from probing a circuit to training a new behaviour around the apparatus. The scientist thinks they are studying memory or fear; the animal thinks it is learning the rules of an unusual world.

When optogenetic results do hold up under this pressure, they become far more than demonstrations. They become anchors for mechanistic understanding: not simply "light changed behaviour," but "a defined perturbation changed a measured physiological pattern in a way that cannot be explained by heat, artifacts, expression stress, or mechanical disruption." That kind of claim is rarer, harder won, and vastly more valuable.

The irony is that the discipline imposed by these pitfalls often leads to more creative science. If a high-duty-cycle protocol risks heating, researchers invent stimulation patterns that are briefer but smarter. If photoelectric artifacts contaminate recordings, engineers develop new electrode materials or geometries. If expression burdens cells, genetic strategies become more subtle. In a field built on light, the strongest results are often those that survive being viewed in the harshest possible illumination.

Chapter 9

Engrams, Fear, and the Machinery of Meaning

Memories do not sit in the brain like files in a cabinet, waiting to be retrieved from a single shelf. Here the reader watches researchers hunt for the fleeting cell ensembles that store experience, replay them with pulses of light, and probe fear circuits closely enough to ask not merely where emotion lives, but how the brain decides what a signal means.

9.1 The Engram Hunt: Giving Memory an Address Without Turning It Into a Place

The wish that every memory might have an address is an old one. Not an *idea* about memory in general, but a very particular sort of wish: that a remembered afternoon, a learned route home, a feared sound, could be pointed to the way you might point to a house on a street. It is the temptation behind phrases like "where memories are stored," as if the brain were a library and the trick were finding the correct shelf.

The historian of science can tell this story as a series of disappointments. At the start of the twentieth century, Richard Semon gave the engram its name and, with it, a promise: that the past leaves a physical trace. Then came a brutally practical question. If a trace is physical, why can't we locate it?

Karl Lashley made that question into a career. In the 1920s through the 1950s, he trained rats to run mazes and then removed pieces of their cortex, looking for the missing bit that would take the memory with it. What he found was not the satisfying “memory spot” but something closer to a kind of stubbornness. The rats often performed worse after larger lesions, but the exact location of damage mattered less than expected. Lashley summarized this as “mass action” and “equipotentiality”: cognitive function seemed to depend on the amount of cortex available and to be distributed broadly across it.

Those conclusions are sometimes repeated as if they proved that memory is everywhere and nowhere at once. That overstates what lesion studies can do. A lesion is a sledgehammer. When performance degrades, it can reflect lost information, but also lost attention, motivation, movement control, or the ability to link what is known to what is required. And when performance survives, it may mean that the key information is distributed, but it may also mean that the animal can solve the problem in a new way, or that the task never demanded the missing piece in the first place. Lesions were invaluable, yet they were shaped by the limits of the questions they could ask.

The modern hunt for the engram does not reject Lashley so much as refine what “address” means. A memory address does not need to be a single coordinate. It can be more like a network address: a sparse set of nodes that, when active in the right relationship to one another, allows the rest of the brain to reconstruct meaning. The point is not that a memory is “in the hippocampus” or “in the cortex,” but that certain groups of neurons participate in that memory in a way that can be tracked and tested.

That shift has two parts: a theory about what memory ought to look like in a nervous system built for resilience, and a set of tools that turn participation into something you can label and manipulate.

Start with the theory. A brain that stored each experience by dedicating a unique, non-overlapping block of neurons would run out of space quickly, and it would be fragile: damage a small region, lose a lifetime of recollection. A brain that stored everything everywhere would have a different problem: interference. If every neuron were involved in every memory, memories would blur, and learning something new would overwrite what came before.

A plausible compromise is sparse, distributed coding. Only a small fraction of available neurons are strongly recruited for a particular experience, and those neurons are embedded in circuits that partially overlap with other experiences. The overlap allows generalization (today's park resembles yesterday's park), while sparsity helps separate similar episodes (this park is safe; that alley is not). In hippocampal circuitry, especially the dentate gyrus, this idea becomes almost intuitive: the dentate gyrus sits at a gateway where it can help turn overlapping inputs into more distinct patterns, a computation often described as pattern separation. If an experience recruits a sparse ensemble there, that ensemble can become a kind of index: not the whole memory, but a handle that can help reactivate a broader constellation of representations across the brain.

None of that, by itself, provides an experimental address. It merely sketches what an address *could* be.

The tool-driven part of the story begins with a simple biological fact: neurons briefly announce their recent activity. When a neuron fires intensely as part of learning, it does not only send electrical signals; it also turns on a genetic program. Immediate early genes such as *c-fos* or *Arc* ramp up their expression within minutes. They are not "memory genes" in any mystical sense. They are closer to a molecular flag that says, “This cell has been active enough to change.” That flag does not stay raised for long, and that is precisely why it is useful. It creates a time window.

With that window, the problem of finding an engram changes shape. Instead of asking, “Where is the memory stored?” you can ask, “Which neurons were strongly active during this particular experience, during this particular hour?” Those neurons become candidates. They are not yet the engram in the full philosophical sense. They are simply a defined population that can be found again.

Finding them again is where genetics becomes more than description and starts to function like logic. The central trick is to link activity-dependent gene expression to an effector gene: a gene that encodes something you can later use to manipulate that same neuron. If the active neuron can be made to express a light-sensitive protein for long enough, it becomes optically addressable. It acquires a tag that is both a record of participation and a handle for intervention.

The genetic engineering often uses promoters and recombinases. A promoter is a stretch of DNA that controls when a gene turns on. If you put your effector gene under a promoter that is active in a particular cell type, you restrict expression to that type. This matters because “memory cells” are not a single class. In a region like the hippocampus, excitatory principal neurons carry much of the long-range signal, while inhibitory interneurons sculpt timing and gating. Mixing them together in an intervention can be informative, but it can also be a recipe for confusion.

Cre-lox recombination is one of the workhorses for making this specific. Cre is an enzyme that recognizes lox sites and can flip or remove DNA segments between them. This lets experimenters build genetic constructs that are silent until Cre acts, and then permanently turned on. It is a biological equivalent of a one-way switch.

Combine those elements and you get a programmable filter: only neurons of the desired type, active in the desired time window, will permanently flip into a tagged state. Days later, the animal can be in a different context, and those same neurons can still be identified and controlled.

There is a practical reason this counts as a breakthrough and not merely a clever label. Brains do not offer tidy boundaries between "encoding neurons" and their neighbors. Electrodes can record activity, but it is hard to know whether a recorded neuron is crucial or merely correlated. Imaging can visualize many cells at once, but early optical imaging often traded spatial coverage for temporal precision and still struggled to reach deep structures without invasive optics. Traditional stimulation methods, like electrical microstimulation, activate whatever fibers and cell bodies happen to be nearby. The old approaches made it easy to argue endlessly about whether a manipulation targeted the right thing.

Activity-dependent tagging plus optogenetics changed the argument from a debate about plausibility into an experiment about causality. The question becomes: if these neurons were active during learning, and if you later reactivate them, can you drive a recall-like state? And if you silence them, do you prevent normal recall? This is where the logic of sufficiency and necessity, introduced earlier, becomes more than a philosopher's distinction. It becomes a set of tests that can be run.

Before the drama of switching memories on and off, there is a quieter conceptual payoff: "address" no longer means "a place in the brain." It means "a population that can be re-accessed." You can have an address that is distributed across regions, cell types, and layers. You can have more than one address for the same event, depending on which features of the event you use as a cue. You can even have an address that changes as memories stabilize over time, without implying that the memory has literally moved the way a file moves from one folder to another.

A concrete example helps. Consider contextual fear conditioning, a standard laboratory paradigm because it is simple, reliable, and ethically bounded. A mouse is placed in a particular chamber with a distinct smell, lighting, and texture. It receives a mild foot shock. Later, when returned to that same chamber, it freezes: a defensive behavior that is easy to quantify. The memory is not only "shock happened." It is "shock happened *here*," with 'here' defined by a bundle of contextual cues. Hippocampal circuits are deeply involved in contextual representation, while the amygdala and related structures help link cues to defensive responses.

In the modern engram framing, the hippocampus does not store fear as an emotion; it provides a structured representation that can be bound to significance. During conditioning, a sparse ensemble in the dentate gyrus becomes strongly active as the context is encoded. Those neurons can be tagged during that learning period using an activity-dependent system. Later, those tagged neurons can be reactivated with light if they express an excitatory opsin such as channelrhodopsin-2.

This does not mean that the tagged dentate gyrus ensemble *is* the fear memory. It means the ensemble is a gateway into a broader circuit state that the animal has learned to treat as meaningful. The dentate gyrus can be the index that points downstream structures toward "that context," and the downstream structures can supply the learned defensive response.

The distinction is not pedantic; it is protective. Without it, it becomes too easy to tell a story in which a memory is a little film clip stored inside a set of cells, waiting to be played. Brains do not work like that. A memory is closer to a pattern of readiness. When the right cues arrive, the brain slips into a coordinated state: certain expectations are primed, certain actions are more likely, certain interpretations of ambiguous sensory input become favored. The engram ensemble is a lever that helps tip the system into that state.

To understand why this is still tricky, imagine trying to "tag the active neurons" during an experience that unfolds over minutes. Some neurons will fire because the animal is exploring; others because it is startled; others because it is processing smell; others because it is moving. The immediate early gene flag does not distinguish "essential encoding" from "incidental activity." It marks strong activation. That is useful, but it is not selective for meaning.

This is where cell-type targeting and careful experimental design do heavy lifting. If tagging is restricted to excitatory neurons in a region thought to carry contextual representation, the tagged set is more likely to reflect the aspect of the experience under study. Timing can be tightened to capture a narrower epoch. Control conditions can separate context exposure from shock delivery, or compare tagging in one environment to behavior in another. These are not bureaucratic details; they are the difference between an address and a crowd.

The deeper caution is that a tagged ensemble is an operational definition: "cells active during a learning episode that can be reactivated later." Operational definitions are the engine of experimental science, but they can be mistaken for the thing itself. A tagged set can be too large, capturing multiple processes. It can be too small, missing crucial participants that were active below threshold. It can be biased toward neurons that are simply more excitable and thus more likely to cross the activation threshold, regardless of what they represent. And tagging can alter the system by adding proteins and changing excitability. Every step that makes the engram tractable also risks changing it.

This is why modern engram work tends to accumulate evidence rather than claim a single decisive proof. Correlation is the starting point: neurons are active during encoding and later during recall. Necessity tests ask whether disrupting those neurons disrupts recall. Sufficiency tests ask whether activating them can substitute, at least partially, for natural cues. Mechanistic evidence asks whether synaptic changes occur within the ensemble and in its downstream partners in a way that can plausibly store information. Each of these criteria can fail for reasons unrelated to memory storage itself, and each can succeed in a way that still leaves room for alternative interpretations. Together, they narrow the plausible stories.

The "address without place" idea also helps with a persistent confusion: the role of the hippocampus in episodic-like memory. People often hear that the hippocampus is "the seat of memory" and then wonder why hippocampal damage does not erase everything. The more accurate view is that hippocampal circuits are crucial for certain kinds of relational and contextual binding, especially when memories are new and rich in detail. With time, many memories become less dependent on hippocampal indexing and more supported by distributed cortical networks. This does not require that the hippocampus hands off a complete copy of a memory to the cortex like a clerk transferring a file. It can mean that the cortex gradually becomes capable of reconstructing the relevant pattern without needing the hippocampal cue, because repeated reactivations have strengthened cortical-cortical links.

Optogenetic access makes this more than a theory, because it allows experimenters to test which ensembles are effective at driving recall at different times after learning, and in which regions. When a memory is young, hippocampal tags may be particularly potent handles. Later, cortical ensembles might become more effective. The "address" of a memory can be time-dependent, and there can be multiple partially redundant addresses that support recall under different conditions.

A final historical anecdote brings the transformation into focus. Lashley, late in his career, reportedly remarked that his results suggested that learning was "not possible." The remark is often quoted with a smile because learning is obviously possible; rats learn mazes, people learn languages, and we do not wake each morning as blank slates. The point was not literal despair. It was a recognition that the simple mapping he sought—this maze memory in that cortical patch—was not emerging from the tools of his time.

What has changed is not that the engram suddenly became simple. It is that the questions became sharp enough to match the brain's actual complexity. The hunt no longer asks for a single pin on a map. It asks for a reproducible way to return to the same sparse ensemble and test what causal role it plays inside a living circuit.

What remains, after the engineering and the diagrams, is a more human point about meaning. A memory is not valuable because it is stored; it is valuable because it can be *used*. The brain's trick is to turn a fleeting pattern of activity into a durable change, and then to treat the reappearance of that pattern—whether triggered by the world or by internal signals—as a reason to expect, to choose, to avoid, to seek. The modern engram hunt earns its excitement not by promising a tiny vault where the past sits untouched, but by offering a way to touch the living circuitry that turns experience into significance.

9.2 Recall by Remote Control: What It Means to Make a Memory Flicker Back

A memory can feel like a private possession: sealed inside the skull, accessible only through the mind's own winding routes. That intuition is part of why optogenetic "recall" experiments landed with such force. The headline version was irresistible: shine light into the brain, and a memory appears. It was described, in public-facing retellings, as "conjuring" the past. The laboratory reality is more modest and, in a way, more unsettling. Light is not a projector. It does not play back a scene. It nudges a living circuit into a state the animal has learned to treat as meaningful, and the animal's brain does the rest.

The drama begins with the basic fact that neurons speak in electrical pulses. As we saw earlier, an action potential is a brief, all-or-nothing spike, and the brain's computations depend heavily on the timing and patterning of these spikes. Channelrhodopsin-2 (ChR2) offered an unusually direct way to enter that language. ChR2 is a light-gated ion channel: when illuminated with blue light, it opens and allows ions to flow, depolarizing the neuron and making spikes more likely. Deliver light in brief pulses and you can drive spikes with millisecond-scale control. That temporal precision is the difference between "activating a region" and pushing a particular set of neurons to fire in a way downstream circuits can actually read.

Optogenetics also changed the feel of causal claims. Older interventions were often blunt. Drugs are slow and diffuse; electrical stimulation can recruit axons passing through the area; lesions are permanent and disruptive. By contrast, optogenetic stimulation can be brief, repeatable, and targeted to genetically specified neurons that were tagged during a particular experience. That gives the experimenter a new kind of sentence to write in the brain: *these* neurons, *now*.

The signature demonstration, popularized by work from Susumu Tonegawa's group, including Xu Liu and Steve Ramirez (2012), used a familiar behavioral assay: contextual fear conditioning. A mouse learns that a particular context predicts a mild shock. Later, in that context, it freezes. Freezing is a convenient readout: it is species-typical, easy to quantify, and tightly linked to defensive circuits. Yet freezing is not "fear" in the everyday sense; it is a measurable motor pattern that often accompanies a defensive internal state.

The critical optogenetic twist was to tag a sparse dentate gyrus ensemble that was active during the learning episode, and later stimulate that ensemble with blue light while the animal sat in a different, neutral environment. When the tagged neurons were reactivated, the mouse froze in the neutral context. The behavioral output fit an intuitive description: something about "the memory" had been brought back.

That observation is a sufficiency test in its purest form. The external world is not providing the original cues. The chamber is not the shock chamber. The smell is wrong, the walls are wrong, the floor is wrong. Yet activating the tagged ensemble is enough to push the animal into a defensive mode. In ordinary language, the animal behaves as if it "remembers" the shock context.

The phrase "as if" is doing important work here. A living brain is not a tape recorder. Recall is not simply replay. Recall is a reconstruction that depends on the brain's current state, its expectations, and the cues it treats as relevant. When optogenetic stimulation elicits freezing, the most conservative statement is that the stimulation is sufficient to produce a recall-like state that normally accompanies retrieval of the fear memory. That is already extraordinary. It is also carefully narrower than the claim that the entire subjective content of the original experience has been reinstated.

There are at least three layers that can be pulled apart.

First is the intervention: patterned blue light delivered to tissue expressing ChR2.

Second is the immediate physiological effect: spiking in the targeted neurons and changes in the dynamics of local circuits (including downstream synaptic release and inhibition).

Third is the inference: because a behavior associated with retrieval is observed, the animal is inferred to be in a retrieval-related internal state.

The gap between the third layer and everyday notions of "remembering" is where misunderstandings tend to breed. That gap is not a flaw. It is the normal space between brain and mind, where neuroscience has to be precise about what is measured.

That middle layer—what stimulation actually does to a circuit—also explains why timing and pattern matter. In the natural brain, a memory trace is not simply "these neurons fire." It is "these neurons fire at particular times, in relation to one another, and in relation to ongoing rhythms and inputs." Artificially driving an ensemble can partially approximate this, but it can also recruit the ensemble in a synchrony that never occurs naturally. The animal may still produce a coherent behavioral response, because downstream circuits can tolerate variability and because the brain is an inference machine: given an internal cue that reliably predicts danger, it chooses a defensive policy.

This helps explain both the power and the limits of remote-control recall. The power lies in the fact that the cue is internal, generated by selectively activating neurons that were active during learning, rather than by the external environment. The limits lie in the reality that the internal cue is crude compared with the richly structured pattern produced by natural recall.

A useful everyday analogy is to think about hearing the opening notes of a song from your teenage years. A few notes can bring back a flood of associations. The notes do not contain the entire past. They are an index that prompts reconstruction. Optogenetic stimulation can function like an artificially generated cue: strong enough to pull the system into a learned attractor state, but not a full sensory reenactment.

The most vivid sign that stimulation is being treated as a cue—and not merely as a forced motor output—came from a related experiment that has become famous for its philosophical provocation. Ramirez and Liu (2013), again in Tonegawa's group, showed that reactivating a tagged context ensemble could be incorporated into new learning in a way that produced a behavioral "false memory."

The logic is simple and eerie. First, tag the ensemble active in context A, a harmless environment. Later, place the mouse in context B and deliver shocks, but at the same time optogenetically reactivate the tagged ensemble for context A. The animal's brain receives an external context (B) and, simultaneously, an internal cue that signals context A. Later still, when returned to context A, the mouse freezes, as if it had been shocked there.

The point is not that the mouse has been "tricked" in a cartoonish way. It is that the brain can treat internally generated patterns of activity as part of the evidence about what is happening, and can bind that evidence to outcomes like shock. Under normal circumstances, the evidence is supplied by senses and by internally generated retrieval processes. The experiment shows that artificially reactivated ensembles can enter that same computational stream. The memory system is not a passive archive. It is a machine for assigning outcomes to predictors, whether those predictors arrive through the eyes or through a fiber optic.

This pushes against a comforting picture in which memories are stable objects that can only be accessed, not altered, by recall. Instead, recall is a form of re-entry into a circuit state, and re-entry creates opportunities for updating. That idea sits behind a great deal of work on reconsolidation and extinction, but the optogenetic "false memory" result makes it unusually concrete. A memory cue, when reactivated, can become a component of a new association. "Remote control" is not just retrieval; it can also be editing, if paired with the right experiences.

The phrase "false memory," though, can mislead if it suggests a human-style narrative error. The mouse is not reporting an autobiographical story. It is expressing a learned defensive response in a particular context. In animal models, memory is often inferred from behavior, and that is both a strength and a constraint. It is a strength because behavior is observable and quantifiable. It is a constraint because multiple internal states can produce similar behaviors, and a single internal state can express itself through many behaviors depending on context.

This is why careful controls are not optional decoration in engram studies. When a mouse freezes during light stimulation, several alternative explanations must be considered. Perhaps the stimulation is aversive by itself, independent of any memory. Perhaps it disrupts movement. Perhaps it alters arousal in a way that biases freezing. Perhaps it activates fibers passing through the region, not only the tagged neurons. These are all addressable concerns, but they do not evaporate merely because optogenetics feels "clean." As we saw in the earlier discussion of the cleanliness illusion, precise tools still need precise reasoning.

The most persuasive studies build a converging case. They show that stimulation of tagged ensembles has a stronger effect than stimulation of untagged neurons, or neurons tagged in a different context. They show that animals behave differently depending on whether the tag was set during the critical learning episode. They often include anatomical verification: the tagged ensemble is sparse, and its projections align with known pathways. Sometimes they compare the downstream activity patterns elicited by natural recall and by optogenetic stimulation, asking whether artificial reactivation partially reinstates the brain-wide signature of retrieval. Recent brain-wide mapping supports the idea that reactivation can partially reproduce naturalistic patterns, but the degree of match remains an empirical hinge, not a guaranteed identity.

There is also a subtler conceptual worry: stimulating an ensemble might cause an effect because the ensemble is genuinely part of the memory trace, or because it is a convenient entry point into the relevant network. Those options can overlap. A dentate gyrus ensemble might be a crucial component of a hippocampal index, and also a particularly powerful way to drive downstream structures. The important scientific move is to avoid confusing "effective handle" with "complete content." A steering wheel controls a car, but it is not the entire engine.

A brief historical detour makes this restraint easier to appreciate. In the middle of the twentieth century, Wilder Penfield electrically stimulated the brains of awake epilepsy patients during surgery and elicited vivid experiences: a song, a voice, a scene. Penfield sometimes described these as "memories." The episodes were compelling enough that they shaped public imagination for decades: press the right spot, and the past returns in full color. Later work complicated that story. Some experiences were fragmentary, some were more like dreams, and stimulation could also produce sensations without any clear autobiographical reference. Penfield's work remains important, but it is now read with a more careful sense of what stimulation can and cannot tell us.

Optogenetic recall experiments sit in that same lineage, with a crucial improvement: they can target neurons selected by their participation in a specific learning event, rather than whatever happens to lie under an electrode. That is why they speak so directly to engrams. Yet the interpretive challenge is similar. When you intervene in a circuit, the brain responds as a dynamic system, not as a filing cabinet.

The dynamic-systems view also clarifies why patterns of stimulation can produce different outcomes. A burst of synchronous spikes may shove a network into a defensive mode, while a weaker or differently timed pattern may fail to cross the threshold needed to recruit downstream structures. Moreover, downstream interpretation is not fixed. The same hippocampal cue can be read differently depending on the state of the amygdala, the prefrontal cortex, neuromodulators such as noradrenaline, and the animal's current goals. A signal is never just a signal; it is a signal *to someone*, in a particular physiological moment.

This is where the language of "replay" needs care. In some contexts, "replay" refers to measurable neural sequences, such as hippocampal place-cell sequences that recur during rest or sleep. That is a technical use of the term, tied to electrophysiology. In everyday speech, replay implies a faithful reenactment. Optogenetic stimulation does not guarantee fidelity. It forces one component of a larger pattern and watches what the organism makes of it.

That observation does not weaken the science. It points to the real prize: understanding how meaning is assembled by circuits. If a memory can be prompted by reactivating a sparse ensemble, then that ensemble is not merely correlated with memory—it is functionally embedded in the machinery that produces memory-guided behavior. If an artificially reactivated ensemble can be bound to a new outcome, then the boundary between retrieval and new learning is more permeable than intuition suggests. The past is not simply accessed; it is re-entered, and re-entry is an opportunity for revision.

For a human reader, the relevance is hard to miss. People who live with trauma often describe intrusive memories that arrive unbidden, as if a cue inside the brain has become too easily triggered. People with depression may find that certain interpretations of the world reassert themselves with stubborn force. Addiction can feel like a learned significance that hijacks attention and choice. Optogenetic engram studies are not direct models of these experiences, and freezing in a mouse is not a confession of terror. Yet the circuit logic is suggestive: internal patterns can act as cues, cues can be strengthened or re-linked, and small ensembles can tip whole networks into states that govern behavior.

Remote-control recall, at its best, is not a parlor trick. It is a disciplined way of asking what the brain treats as evidence, and how much of that evidence can be carried by a sparse, reactivatable group of neurons. The light is only the opening move; the meaning is what the circuit decides to do with the activity it has been handed.

9.3 The Amygdala's If-Then Rules: Fear as a Circuit Decision, Not a Blob of Emotion

The word "fear" does double duty. It can name an experience—what it feels like to anticipate harm—and it can name a suite of defensive actions—freezing, fleeing, a racing heart, a surge of attention toward threat. Everyday speech blurs these together, and the blur seeps into brain talk. The amygdala becomes "the fear center," as if a small almond-shaped structure were a tank where fear is stored and from which it spills.

As we saw, that picture does not survive contact with the circuitry. The amygdala is better treated as a hub in a distributed set of threat-related circuits: the hippocampus contributes contextual representation; medial prefrontal cortex (mPFC) contributes regulation and extinction-related control; downstream hypothalamus and brainstem can implement defensive outputs. Within the amygdala itself, different nuclei and local inhibitory microcircuits support distinct computations—association, expression, inhibition—so that "fear" begins to resemble a set of conditional rules rather than a single emotion blob.

Those rules are often learned in a laboratory with almost comical simplicity. A tone sounds; a mild shock follows. Later the tone alone can elicit freezing. The power of the paradigm is not its realism. It is its clarity: the animal's brain must learn an *if-then* relationship that can be measured. If the tone predicts danger, then freeze. If the context signals safety, then do not. If a safety memory has been learned through extinction training, then the same tone can be handled differently depending on where and when it appears.

The amygdala's centrality in this learning has been recognized for decades, especially through the work of Joseph LeDoux and others who mapped sensory inputs to amygdala-related circuits, and through the synthesis of Stephen Maren (2001), who helped clarify how associative learning, plasticity, and defensive behavior fit together. Yet those foundational maps did not settle an important question: what exactly is the amygdala *doing* at each stage? Is it where the association is stored? Is it where the behavior is selected? Is it a relay whose job changes across time? With lesions, one could argue about necessity, but it was difficult to specify which cells and which pathways were responsible, and at which moment of the learning and recall process.

Optogenetics made the timing and routing explicit. Instead of asking whether the amygdala is involved, researchers could ask what happens if one input is perturbed during learning but left alone during recall, or if an output pathway is briefly silenced only when the animal is about to freeze. The result is a more operational view of fear conditioning: not "amygdala on equals fear," but "this input arriving at this time drives plasticity; this output pathway drives expression; this inhibitory gate determines whether expression is permitted."

A useful way to keep this straight is to start with a plain-language account of the task, and then translate it into circuit terms.

Imagine a mouse in a chamber. A tone comes on, and a second later there is a shock. On the first trial, the tone is not frightening in itself; it is just sound. But it becomes informative, and the information is about the future. The tone becomes a predictor. Prediction is the substrate of an if-then rule.

In circuit terms, the tone is represented in auditory pathways and delivered to amygdala-related circuits. The shock is represented in somatosensory pathways and also delivered. A place where those streams converge is a place where synapses can change in a way that encodes the predictive relationship: "tone $\rightarrow$ shock." Much of the classic account places this convergence and plasticity in the lateral amygdala and basolateral amygdala (often grouped as BLA). But the BLA is not a single pool. It contains diverse neuron classes with different projection targets, and it is threaded with inhibitory interneurons that sculpt which principal neurons fire, when, and with what reliability. In other words, the BLA is not merely a storage bin for associations; it is a computation.

From there, defensive actions are expressed through downstream routes. The central amygdala (CeA) and its connections to hypothalamus and brainstem are often emphasized as output controllers: pathways that can influence freezing, autonomic arousal, stress hormone release, and orienting. Again, the idea is not that CeA "contains fear." It contains microcircuits that can route learned significance into action.

The microcircuits are where the *if-then* quality becomes hard to ignore. Inhibitory neurons can act as gates: they can suppress a downstream output unless certain conditions are met. Such gating is not an abstract metaphor; it is built into the architecture. Within the amygdala, inhibitory microcircuits can implement conditional control—if context predicts safety, then suppress output—even while the tone is still recognized as relevant. That separation between recognizing a cue and choosing a response is one reason fear conditioning is such a rich model: it forces the brain to bind prediction to behavior, but it also leaves room for regulation.

Regulation becomes particularly visible during extinction. Extinction training is what happens when the tone is presented many times without the shock. The freezing response gradually diminishes. Anyone who has tried to unlearn a phobia, or who has undergone exposure therapy, will recognize the basic pattern: reduction in defensive response with repeated safe exposure. The brain, however, has learned two things: the tone can predict danger, and the tone can be safe under certain circumstances. That is why extinction is often framed as new learning that suppresses expression rather than deletion of the original association.

Here the mPFC enters as more than a vague "control center." Extinction retrieval depends on prefrontal control: inhibiting infralimbic cortex during retrieval can impair extinction expression, increasing freezing. That is a strikingly circuit-level statement. It says that when the animal is tested after extinction, the ability to *express* safety depends on activity in a specific prefrontal area at a specific moment. The original association can still be present; what changes is whether it is allowed to drive behavior.

Optogenetics is particularly well suited to making this point because it can be used in a temporally bounded way. A pharmacological inactivation of mPFC can last for minutes or hours, muddying the interpretation by affecting attention, arousal, or movement throughout the session. A brief optogenetic inhibition can be timed to the moments when the tone is played. The animal can explore normally, then, at the critical time, the prefrontal gate is temporarily shut. Freezing rises. The animal is not forced to freeze by direct stimulation of motor outputs; rather, a control signal that would normally suppress a defensive policy is removed. The if-then rule is revealed by subtracting a gate.

That story also clarifies why fear conditioning cannot be reduced to a single anatomical site. The hippocampus is essential in contextual fear conditioning because it helps represent "where" and "when" in a way that can be bound to outcomes. Context is not merely background; it is part of the conditional. If the animal learns "the tone is dangerous in this context but safe in that one," then hippocampal signals about context can determine whether amygdala-driven outputs are permitted. The same cue can be read as dangerous or safe depending on a broader network state. In real life this is the difference between a backfire on a film set and a backfire in a war zone. The auditory stimulus can be nearly identical; the meaning is not.

Circuit-based work has also made a subtler unit of analysis feel natural: sometimes the decisive thing is not a region, but a connection. Projection-defined pathways—such as BLA $\leftrightarrow$ mPFC—can be causal levers for fear expression and updating. That is a deep change in how the brain is described. Saying "the amygdala controls fear" implies a monolithic source. Saying "the BLA-to-mPFC pathway supports this form of regulation" implies routing, directionality, and the possibility that the same region participates in multiple, even opposing, functions depending on which projection is engaged.

This projection logic also helps resolve an apparent paradox that often confuses non-specialists. If the amygdala is so central, why can fear be both stubborn and flexible? Why does a learned fear sometimes generalize too widely, and at other times become exquisitely context-specific? The answer is that the association and the expression are separable, and they are separable because the circuit has multiple junctions at which conditions can be applied. Plasticity in the BLA can encode that a cue predicts danger. Inhibitory gating in the amygdala and top-down signals from mPFC can decide whether the prediction is acted upon here and now. Hippocampal context signals can bias that decision. Downstream brainstem and hypothalamus can implement different flavors of defensive behavior. The overall behavior is the end of a chain of *if-then* operations, each of which can be adjusted.

A historical anecdote makes the importance of timing especially vivid. Ivan Pavlov's dogs were not just salivating machines; they were timing machines. Pavlov noticed that the interval between a bell and food shaped learning. When the cue and outcome were properly ordered and close in time, conditioning was strong; when they were misaligned, it weakened. The nervous system learns contingencies that have temporal structure. Modern fear conditioning is Pavlov's insight with a sharper knife: instead of only measuring what the animal does, one can perturb a pathway at acquisition, consolidation, or retrieval and see which part of the if-then computation is affected.

Timing has another consequence that optogenetics exposes: the same circuit element can serve opposite roles depending on when it is engaged. Activity in a pathway during acquisition can drive plasticity;

activity in the same pathway during retrieval can drive expression; inhibition at one time can prevent learning while inhibition at another time can prevent behavior. Without temporal precision, these roles get conflated, feeding the misleading sense that the amygdala is simply "active during fear." With temporal precision, the amygdala starts to resemble a programmable switchboard.

This framing also helps in thinking carefully about what behavioral readouts mean. Freezing is a defensible measure of a defensive state in rodents, but it is still a motor output shaped by many influences. An animal can fail to freeze because it is less threatened, but also because it is distracted, sedated, or committed to another defensive strategy. Conversely, it can freeze more because the threat memory is stronger, or because a safety gate is weakened, or because arousal systems have been pushed toward vigilance. Optogenetic perturbations can be exquisitely precise at the level of neurons, yet still produce changes that must be interpreted in behavioral context.

The most mature lesson of the amygdala's if-then rules is that "meaning" is not painted onto a stimulus by a single emotional organ. Meaning is the conditional relationship between cues, contexts, internal state, and predicted outcomes, implemented by pathways that can be strengthened, gated, and rerouted. A tone can be "danger" when the prefrontal brake is off and the context matches, and "nothing" when the brake is on and the context signals safety. The amygdala participates in both, not as a bottle of fear, but as a set of nodes where prediction is translated into action and where inhibition can decide that prediction will not rule the moment.

That circuit view can sound cold, but it captures something humane. Fear is not only a feeling; it is a negotiation between protection and possibility. The brain must weigh what has happened before against what is happening now, and it must do so quickly. A system built from if-then rules—learned, revisable, and subject to control—can fail in recognizable ways. It can gate too little, producing persistent defensive responses that outlive their usefulness. It can gate too much, blunting appropriate caution. It can misread context. It can treat internal cues as external certainty. The promise of circuit-level understanding is not that it drains fear of meaning, but that it identifies where meaning is made operational: at the junctions where a prediction becomes a choice.

9.4 Editing Fear Without Erasing the Brain: Silence, Control, and the Problem of Overclaiming

Turning neurons on with light is the demonstration that grabs attention; turning them off is where the interpretive discipline begins. Activation experiments tempt us with a vivid cause-and-effect story: press the neural button, and a behavior appears. Inhibition experiments sound even closer to therapy: silence the troublesome circuit, and the fear goes away. Yet silencing is rarely a clean subtraction. It is an intervention into a living system that compensates, rebounds, reroutes, and sometimes simply fails to express what it still "knows."

The underlying hope is easy to empathize with. Fear can be lifesaving, but it can also become tyrannical. Phobias, panic disorder, post-traumatic stress disorder, and some forms of anxiety are marked by defensive responses that are too strong, too easily triggered, or too hard to turn off. If the brain encodes "danger" in circuits, perhaps circuits can be adjusted. Optogenetics seems, from the outside, like a surgical instrument for meaning.

The scientific reality is more cautious: optogenetics is an instrument for testing causal roles within circuits, and any leap from a rodent's freezing behavior to a human's lived fear must be argued, not assumed. The two-system framing—one set of circuits for defensive responses and another for the conscious experience we label fear—serves as a useful guardrail. Manipulating defensive behavior circuits can be transformative for understanding and, one day, for treatment, but it does not automatically imply direct access to subjective fear.

Inhibitory optogenetics begins with opsins that do the opposite of channelrhodopsin-2. Instead of opening a channel that depolarizes a neuron, inhibitory opsins make spiking less likely. Halorhodopsin and Arch are classic examples. They are light-driven proteins that, through ion movements, tend to hyperpolarize neurons or otherwise suppress activity. In practice, what they "silence" depends on the cell, the expression level, the light intensity, and the moment-to-moment state of the circuit. The promise is temporal precision: inhibition can be delivered during a narrowly chosen behavioral window—during a tone, during a retrieval test, during the seconds when a shock is expected—rather than for an entire session.

That temporal precision is crucial because fear learning has stages with different logic. Acquisition is when associations are formed. Consolidation is when they stabilize. Retrieval is when the learned relationship is expressed or used. Extinction retrieval adds another layer: a prefrontal gate that can suppress expression under safe conditions. Silencing the "wrong" pathway at the "wrong" time can produce an effect that sounds like memory erasure but is actually a temporary failure of expression.

A simple thought experiment shows why necessity tests are delicate. Suppose a mouse normally freezes when it hears a conditioned tone. Now suppose you inhibit a pathway and the mouse freezes less. It is tempting to say the pathway is "necessary for fear." But freezing is a behavior, not a self-report. Reduced freezing could mean the animal is less threatened, or it could mean it is distracted, less able to hold still, less attentive to the tone, or shifted toward a different defensive strategy. A well-designed experiment tries to rule these out, but the logical point remains: necessity claims live at the intersection of circuit manipulation and behavioral interpretation.

Optogenetic inhibition introduces its own technical confounds. Delivering light into tissue can cause heating; heating can change neural excitability and blood flow. Inhibiting neurons can produce rebound firing when the light turns off, potentially creating a burst of activity that the experimenter did not intend. Light delivery hardware can alter behavior by tethering the animal. Stimulation can be perceived indirectly if it changes sensory cues or induces discomfort. These are solvable problems, but only if treated as central rather than as afterthoughts.

One reason the problem is harder than it seems is that brains are not engineered to fail quietly. If one pathway is suppressed, other pathways may compensate, sometimes within seconds. A defensive response might be reduced not because the memory is gone, but because the usual route for expressing it has been blocked and the alternative route is slower or requires stronger input. When the inhibition ends, expression can return. The memory did not vanish; it was temporarily unable to control action.

This is why claims of "erasure" are intrinsically hard to justify. Engrams are distributed and time-evolving, so silencing one node or one ensemble may change behavior without eliminating the underlying representation. A memory that can be driven by one handle today may be driven by another handle tomorrow. A circuit can be quieted at one junction and still persist in a broader network state.

The story of reconsolidation shows how the field came to believe that fear memories might be modifiable in a deeper sense than mere suppression. The core idea is that retrieval can make a consolidated memory temporarily labile, opening a window during which the memory must be restabilized. Reconsolidation work supports a retrieval-dependent lability window in which fear memories can be disrupted or updated via protein synthesis-dependent mechanisms in the amygdala. A classic manipulation in this tradition involved anisomycin, a protein synthesis inhibitor, delivered after retrieval. The conceptual pivot was radical: recalling a memory is not just reading it; it can be a necessary step in maintaining it.

Even here, caution is warranted. Boundary conditions and replication variability exist across paradigms, and the exact conditions under which reconsolidation can be reliably engaged are still debated. Yet the foundational implication remains: there may be times when fear memories are particularly susceptible to modification, and those times are linked to retrieval.

Reconsolidation can be misunderstood as a "delete button." It is not. Blocking protein synthesis after retrieval does not reach into the past and remove an experience. It interferes with molecular processes required for restabilization, and the behavioral effects depend on many factors, including the strength and age of the memory and the details of the retrieval cue. Moreover, a reduction in freezing after such an intervention still raises the same interpretive questions: is the association weakened, or is expression altered, or is retrieval impaired?

The contrast with extinction is clarifying. Extinction is often framed as new learning that suppresses expression. Reconsolidation implies modification of the original trace. Those are not merely theoretical labels; they imply different circuit mechanisms. Extinction retrieval depends on prefrontal control, and optogenetic studies of prefrontal control highlight suppression and regulation rather than literal deletion. When infralimbic cortex is inhibited during extinction retrieval, freezing can increase: the safety gate fails, and the older association regains behavioral control. That is not the return of a "deleted" memory. It is the unmasking of a still-available association when control is removed.

From the perspective of therapy, this distinction is not academic. Exposure-based treatments are closer in spirit to extinction: building a safety memory and strengthening the ability to retrieve it under real-world conditions. Reconsolidation-based approaches, when they work, suggest something closer to updating the original fear memory itself. Both can reduce suffering; they rely on different levers.

Optogenetics adds another provocative lever: changing the emotional value linked to a memory representation. Engram valence-switch experiments, associated with Ramirez and Liu (2014), demonstrated reassignment of emotional value to an engram via pairing reactivation with opposite-valence experiences. In plain terms, an ensemble tagged during a fear-related experience can later be reactivated while the animal experiences something rewarding, and the animal's subsequent behavior suggests that the meaning of that representation has shifted. The memory content—the "where" and "what"—can remain, while the valence—the "good" or "bad"—is re-linked.

This kind of result is often described, in simplified narratives, as "editing memory." The phrase is not entirely wrong, but it needs careful handling. What is being altered is not a detailed autobiographical record. It is the learned linkage between a representation and an emotional or motivational outcome. That linkage is exactly what makes fear conditioning a model for meaning: cues come to *stand for* something. Changing what they stand for changes behavior.

Yet valence switching also highlights why claims must be calibrated. Changing an animal's preference or avoidance tells you something about how the circuit now routes behavior, not about whether the animal's conscious experience has been rewritten. The two-system framing again keeps the language honest. In rodents, one can speak confidently about defensive responses and their control. One can speak more cautiously about internal states inferred from behavior. One cannot, from freezing alone, conclude that a subjective feeling has been erased.

A helpful way to keep one's footing is to imagine a ladder of claims, where each rung demands extra evidence. At the bottom is the intervention: a specific population of neurons or a projection is inhibited during a defined time window. The next rung is neural effect: spiking or synaptic output is reduced as intended. The next rung is behavioral effect: freezing decreases, avoidance changes, exploration increases. The next rung is inferred memory state: retrieval has been disrupted, or the association has been weakened, or safety control has been strengthened. At the top is subjective experience: fear has been reduced as a feeling. The lower rungs are where optogenetics is strongest. The top rung is where translations become speculative unless accompanied by converging evidence in humans.

The need for humility is not a plea for timidity; it is a recognition of how many ways a brain can change behavior without changing what it represents. Consider a familiar human scenario. A person with a fear of flying can be prevented from panicking by taking a sedative. The sedative changes behavior and experience in the short term, but it does not necessarily change the learned association between flying and danger. Alternatively, a person can learn coping strategies that reduce panic while the underlying fear remains accessible. Or the fear can genuinely weaken after repeated safe flights. Similar ambiguities exist in rodents: less freezing may reflect sedation, distraction, enhanced control, altered valuation, or weakened association.

That is why optogenetic "silencing" experiments often focus on tight time windows and pathway specificity. If inhibition is delivered only during retrieval of a cue, and basic locomotion, sensory processing, and arousal are intact, the case that the pathway supports retrieval or expression becomes stronger. If inhibition is delivered during acquisition and later freezing is reduced without general learning deficits, the case that the pathway supports learning-related plasticity strengthens. If inhibition during extinction retrieval increases freezing, the case for top-down control strengthens. Each is a precise statement about an if-then computation, not a general statement that fear has been excised.

Even the technical details can have conceptual consequences. Rebound firing, for example, is not just an artifact; it is a reminder that neurons and circuits have dynamics. A prolonged hyperpolarization can alter ion concentrations and channel states, and when the inhibition ends, neurons may fire more. If that rebound occurs at a moment when the animal is sampling cues, it can inadvertently strengthen associations or trigger behaviors. Heating effects likewise remind us that "light" is not purely informational in biological tissue; it is also energy. These realities do not invalidate optogenetic inhibition. They simply demand controls that match the ambition of the claims.

The most responsible stance toward "editing fear" is therefore neither cynicism nor hype. The promise is real: one can manipulate circuits with a precision that older methods could not approach, and one can begin to tease apart learning, expression, and regulation in the amygdala-centered network. One can even re-link representations to new valences, showing that meaning is not fixed but constructed and updateable. The discipline is equally real: a striking behavioral change is the beginning of an explanation, not its endpoint.

Fear is often described as primitive, but the circuitry that governs it is sophisticated in a very practical way. It must keep an animal alive in a world where cues are ambiguous, contexts shift, and yesterday's danger can become today's harmless noise. A system built for that task will not offer a simple "erase" function. It will offer gates, alternate routes, competing memories, and opportunities for updating. When optogenetics quiets a pathway and behavior changes, the most interesting question is rarely whether fear has been removed. It is which conditional rule has been altered: whether the cue still predicts danger, whether safety control can be retrieved, whether a representation has been assigned a new value, or whether the animal has simply lost a way to express what it still anticipates.

Chapter 10

Dopamine's Doppelgängers: Pleasure, Learning, and Compulsion

Dopamine is often cast as the brain's pleasure chemical, yet the real story is stranger and more revealing: the same circuit can stamp in a lesson, amplify the pull of a cue, energize pursuit, and help trap behavior into habit. By flashing specific neurons and silencing selected pathways at precise moments, this chapter lets the reader watch 're-ward' split into its hidden doubles—wanting, learning, movement, and compulsion.

10.1 The Millisecond Teacher

Pleasure feels like a compass. You taste something sweet, warmth spreads through the body, and the mind files it under *do that again*. It is an easy story to tell about animals and about ourselves: rewards pull us toward them because they feel good, and the brain keeps a running tally of what produced that feeling. Dopamine became the starring chemical in that story partly because its pathways run through places long associated with reward and motivation, and partly because drugs that strongly alter dopamine can so powerfully reshape what people pursue.

Yet the closer scientists listened to dopamine neurons, the more dopamine started to sound less like a warm bath and more like a tap on the shoulder.

The classic listening was done with electrodes, thin wires that can pick up the electrical spikes of single neurons. In the 1990s, Wolfram Schultz and colleagues recorded from dopamine neurons in the midbrain while monkeys learned simple associations: a cue predicts juice; the juice arrives, or it does not. Early in learning, when the juice came unexpectedly, dopamine neurons fired a brief burst of spikes. After learning, when the cue reliably predicted juice, the burst moved earlier in time: it now came at the cue, not at the juice. And if the cue appeared but the juice failed to show up, the neurons paused below their usual firing rate at the moment when juice should have arrived (Schultz, Dayan, and Montague, 1997).

That pattern is strange if dopamine is simply *pleasure*. When learning is complete, the juice is still pleasurable; the monkey still likes it. But the dopamine burst, the dramatic part of the signal, has shifted away from the moment of consumption. The signal behaves less like a readout of enjoyment and more like a message about surprise: *better than expected*, *as expected*, or *worse than expected*. In computational language, it resembles a *reward prediction error*: the difference between what you got and what you predicted.

Prediction error sounds abstract, but it fits a very practical problem. An animal has many senses and many possible actions. The world is noisy, and good outcomes are rare. If you want to learn efficiently, you should pay extra attention to moments when the world updates your beliefs: when a berry bush has fruit after weeks of emptiness;

when a rustle that usually means wind turns out to be prey; when a familiar alley suddenly contains danger. A brain that can broadcast "update your expectations right now" will learn faster than a brain that simply bathes itself in pleasure.

This is where time becomes the main character. Dopamine neurons do not merely raise and lower dopamine levels like a thermostat. They can fire in at least two broad modes. There is *tonic* activity: a baseline rhythm that sets a general level of dopamine release, rising and falling over seconds to minutes. Then there is *phasic* activity: short bursts of spikes that last tens to hundreds of milliseconds. Those bursts are not just smaller versions of tonic changes; they can have distinct effects on synapses and on downstream circuits. A brief pulse can arrive like a punctuation mark in an ongoing sentence of neural activity, telling other neurons: *the thing that just happened deserves extra weight in memory.*

The idea that dopamine teaches through precisely timed pulses was persuasive, but for years it rested on a familiar scientific trap: correlation. Dopamine bursts happened when learning should happen, but perhaps they were merely passengers. Maybe some other set of neurons computed the teaching signal, and dopamine neurons were simply echoing it. Or perhaps the burst did not teach at all, but instead altered movement, attention, or arousal in a way that made learning *appear* stronger. To turn the idea into something firmer, neuroscientists needed a tool that could do something both delicate and bold: trigger dopamine neuron bursts at will, with millisecond precision, and observe what that *causes*.

Optogenetics provided that tool. Channelrhodopsin-2 (ChR2) is a light-gated ion channel: when illuminated with the right wavelength, it opens and allows positively charged ions to enter the neuron, pushing the cell toward firing an action potential. By putting the ChR2 gene into a specific cell type—here, dopamine neurons in the midbrain—and delivering brief flashes of light, experimenters can induce patterns of spikes that resemble natural phasic bursts. If a burst is truly a teaching signal, then an artificial burst, delivered at the right moment, should be able to teach even in the absence of a natural reward surprise. That is a *sufficiency* test: is the dopamine burst sufficient, on its own, to produce learning?

The simplest form of this sufficiency logic asks a blunt question: will an animal work to get a dopamine burst? In 2009, Hsing-Chen Tsai, Feng Zhang, and colleagues showed that optogenetic stimulation of dopamine neurons could support behavioral conditioning: brief phasic patterns were particularly effective at driving learning-related changes (Tsai et al., 2009). In 2011, Ilana Witten, Antoine Adamantidis, and their colleagues demonstrated that mice would perform an action to obtain brief optogenetic activation of dopamine neurons—a kind of self-stimulation, but now targeted to a defined cell type rather than broad electrical stimulation that excites passing fibers (Witten et al., 2011; Adamantidis et al., 2011). The mice were not pressing a lever for sugar water or food; they were pressing for a precisely timed pattern of activity in their own dopamine cells.

At first glance, that might seem to restore the old story: dopamine bursts must be pleasure, because animals seek them. But the deeper lesson comes from the *timing* and *contingency*. Animals do not merely receive dopamine and feel good; they learn what predicts dopamine and what actions produce it. If the burst is delivered regardless of what the animal does, it does not build the same structured behavior. If it is delivered reliably after a specific action, the animal repeats that action. If it is paired with a cue, the cue gains influence. Dopamine is acting less like the reward itself and more like a credit-assignment signal: a way for the brain to tag recent events as causally relevant.

Credit assignment is the everyday problem of figuring out what deserves praise or blame. Imagine you try a new recipe and the dish turns out delicious. Was it the brand of tomatoes, the longer simmer time, the extra salt, or the fact that you were starving? If the brain simply recorded *"that was good"* without specifying what preceded the good outcome, it would not know what to repeat. A phasic dopamine burst, arriving shortly after a cue or action, can help stamp that cue or action with importance, strengthening synapses that were active in the moments just before the burst. In many parts of the brain, synapses are especially modifiable when presynaptic and postsynaptic neurons are active close together in time. Dopamine can act as a third ingredient, gating whether that co-activity produces lasting change. In that sense, dopamine bursts can function like a teacher who does not give you the answer, but tells you which part of your work deserves to be kept.

A particularly elegant demonstration that dopamine bursts can substitute for a prediction error came from Elizabeth Steinberg and colleagues in 2013. They used a learning situation known as *unblocking*. In blocking, an animal learns that Cue A predicts reward. Later, Cue A is presented together with a new Cue B, and the same reward follows. Even though B is present, the animal learns little about it, because the reward is already fully predicted by A. Learning about B is "blocked" by the lack of surprise. Unblocking occurs when the reward becomes unexpectedly better (for example, larger), creating a positive prediction error that allows B to be learned.

Steinberg and colleagues arranged things so that the reward itself did not change, but at the moment of reward delivery they added a brief optogenetic activation of dopamine neurons (Steinberg et al., 2013). The stimulation was timed to mimic a positive prediction error: an extra "better than expected" signal layered onto an otherwise expected reward. The result was striking: animals learned about the new cue in a way that resembled unblocking. Even more telling, the learning persisted after the stimulation ended, consistent with the idea that the burst had driven plasticity rather than temporarily boosting performance. This was not just making animals more active or more eager in the moment; it was changing what they had learned about the world.

The insistence on milliseconds is not pedantry. Learning rules in the brain are often sensitive to tight windows of time. If the teacher's signal arrives too late, the brain may reinforce the wrong thing. Anyone who has ever accidentally trained a pet knows this: praise delivered a few seconds after a dog sits can end up reinforcing whatever the dog did in the meantime. Dopamine bursts are brief enough to avoid some of that confusion. They can arrive like a timestamped stamp of approval, strengthening what was just active rather than what happens to be active seconds later.

That temporal sharpness also helps explain why "dopamine equals pleasure" repeatedly fails as a complete account. Pleasure is typically slower, broader, and more diffuse. The conscious enjoyment of food rises and falls over seconds; the satisfaction after solving a problem can linger. Dopamine bursts, by contrast, can be as quick as a camera flash. A flash can change what a photograph captures without being the scenery itself. In the same way, dopamine can change what the brain captures in memory without being the feeling that memory is about.

This distinction becomes clearer when considering the mismatch between *liking* and *learning*. You can learn from an unpleasant experience with terrifying speed. Touch a hot stove once, and you rarely need a second lesson. You can also enjoy something and still not learn much from it if it is not informative: think of background music you like but stop noticing. A teaching signal should track informativeness, not just enjoyment. The Schultz pattern does exactly that: it peaks when outcomes are informative, either because they were surprising or because they update predictions about the future.

The brain's teaching signals also need to be economical. There are countless cues in an environment: light patterns, smells, textures, internal states. Most of them are irrelevant for predicting reward. A prediction-error-like dopamine burst allows learning to focus on cues that actually reduce uncertainty. When a cue reliably predicts reward, the reward itself becomes less informative, and the teacher shifts its attention to the cue. When a cue fails, a dip in dopamine can help weaken the expectation. This is not an emotional judgment so much as a bookkeeping operation carried out with biological hardware.

Optogenetics adds another layer to this story: it reveals that the exact *pattern* of stimulation can matter. A prolonged, slow increase in dopamine neuron activity is not equivalent to a brief burst, even if the total number of spikes is similar. Phasic patterns are particularly potent for reinforcing discrete actions and for stamping in associations. That makes sense if downstream synapses are tuned to treat quick dopamine transients as a special kind of event: not a general motivational climate, but a precise tag for plasticity.

At the same time, the sufficiency experiments also invite a caution. If a mouse works for optogenetic dopamine stimulation, what is it seeking? It might be seeking a subjective pleasant feeling. It might be seeking a sense of energized pursuit. Or it might be seeking the internal confirmation that something it just did was "correct," even if that confirmation is not consciously accessible. Animals will work for many kinds of brain stimulation that do not map neatly onto human pleasure. The power of optogenetics is that it can show what a signal can *do* in a circuit. It does not automatically tell us what that signal *feels like* from the inside.

A useful way to keep this straight is to treat dopamine bursts as messages in a learning system rather than as the reward itself. Rewards are things like food, warmth, safety, social contact, and sexual opportunity: events that matter for survival and reproduction. Pleasure is one way the brain represents the value of those events. Dopamine bursts are closer to an internal annotation: "this outcome was better (or worse) than expected; update your model and your policy." You can annotate without savoring. You can also savor without needing to annotate, once the world is predictable.

This framing helps with a subtle but common confusion: the temptation to assume that if something reinforces behavior, it must be pleasurable. Reinforcement, in the strict behavioral sense, means that a consequence increases the probability of the behavior that produced it. A consequence can be reinforcing because it feels good, but it can also be reinforcing because it reduces uncertainty, relieves a drive state, or triggers a powerful teaching signal that strengthens certain synapses. In other words, "the animal will do it again" does not necessarily imply "it enjoyed it in a way comparable to eating sugar."

The millisecond teacher also makes sense of why addiction is so insidious. Many addictive drugs cause large dopamine increases, sometimes in patterns that mimic or amplify phasic transients. If dopamine bursts are teaching signals, then drug-induced dopamine is not merely pleasure poured into the brain. It is also potentially a stream of lessons: cues, places, stress states, and rituals surrounding the drug get stamped with importance. The brain learns not only that the drug is valuable, but that the predictors and precursors of the drug are valuable too. A doorknob, a street corner, a particular friend's text message, a particular kind of boredom can become instructive events, because the drug has repeatedly delivered a teacher's stamp in their presence.

None of this diminishes the reality of pleasure. It refines its role. Pleasure can motivate and guide, but learning requires something more surgical: a signal that arrives at the right time, aimed at the right synapses, carrying information about prediction. Dopamine bursts are one of the brain's ways of doing that surgery, and optogenetics made it possible to test the idea with a precision that older tools could not match.

A teacher's usefulness is measured by how well it assigns credit. The remarkable thing about dopamine is that it can assign credit at the speed of thought, shaping which fleeting perceptions and actions get written into future tendencies. That is a kind of influence that pleasure alone, however real, cannot easily provide.

10.2 The Cue That Steals the Steering Wheel

A cue can be as ordinary as a lever, a tone, or the soft glow of a screen. At first it is just scenery. Later it can feel like a command.

That is the unsettling twist that addiction research keeps circling back to: the brain does not merely learn what is rewarding, it learns what *predicts* reward, and then it sometimes begins to treat the predictor as if it were the prize. A smell on a jacket, the first sip of a drink, the particular street where a dealer stands—these can become motivational magnets. The cue is no longer a signpost pointing to value; it becomes part of what is valued. People describe this as being "triggered," as if a hand has reached inside the mind to pull a lever. Laboratory animals, for their part, begin to behave in ways that are hard to describe as calm, goal-directed choice. They get pulled.

Earlier, a brief burst of dopamine activity served as a millisecond-scale teacher. The cue story takes that teaching signal and asks a more clinically relevant question: once a cue has gained this power, what keeps it powerful? Which wiring turns a neutral signal into something that can seize the controls of behavior?

Answering that requires a change in experimental posture. Turning dopamine cells on and seeing behavior change is a *sufficiency* argument: this signal can, by itself, stamp in learning or reinforce action. The harder argument is *necessity*: is a particular signal required at the moment a cue appears, or can the brain route around it? Necessity tests are where optogenetics becomes less like a loudspeaker and more like a scalpel. Instead of adding activity, researchers briefly subtract it—and they do so not everywhere at once, but along a chosen path.

To make sense of the cue's grip, it helps to start with a behavioral observation that sounds almost like personality. In a classic setup, a rat hears or sees a cue (often a lever extends into the cage) that predicts that a food pellet will soon arrive in a nearby food cup. With repeated pairings, many rats learn the prediction. But they do not all behave the same way. Some become *goal-trackers*: when the lever appears, they hurry to the food cup, waiting at the spot where the pellet will drop. Others become *sign-trackers*: the lever itself becomes irresistible. They approach it, nibble it, paw it, almost as if it were edible. The lever is not the food; it is a clue. Yet for sign-trackers the clue becomes an object of desire.

This split became a powerful lens through work by Flagel and colleagues (2011) and by Saunders and Robinson (2012). Both groups helped establish that dopamine's role is especially strong when a cue acquires *incentive salience*—the quality that makes it attractive and able to energize pursuit. That phrase can sound like jargon, but it names an experience many people recognize: knowing something is predictive is not the same as feeling drawn to it. A person can know that a notification sound predicts a message without feeling compelled to check. Or they can hear it and feel their attention snap as if yanked by a hook.

The point is not that goal-trackers are "smarter" and sign-trackers are "addicted." Both learned the association. The point is that cue-driven motivation comes in flavors, and dopamine is more entangled with some flavors than others. Addiction becomes less mysterious when it is treated not as a single amplified reward signal, but as an argument among circuits: some circuits put a price tag on predictors, some translate price tags into action vigor, and some provide brakes. In some individuals, and in some drug states, the "cue value" voice wins.

How does one show that a circuit is doing more than tagging a cue as predictive? A particularly telling method is Pavlovian-to-instrumental transfer, usually shortened to *PIT*. In a PIT experiment, an animal first learns a Pavlovian association: a tone predicts sucrose. Separately, it learns an instrumental action: pressing a lever earns sucrose. Then the test comes: during a period when lever pressing no longer produces sucrose (so the animal is not being freshly rewarded for pressing), the tone is played. If the tone increases lever pressing anyway, the cue is not merely informing the animal; it is *invigorating* the action. The tone has become an opportunity signal: *now is a good time to work.*

Wassum and colleagues (2011) linked this cue-triggered invigoration to dopamine transients in the nucleus accumbens, a key node in the brain's reward and motivation circuitry. When the Pavlovian cue appeared, dopamine release in this region rose in a cue-locked way, tracking the cue's ability to energize seeking. That kind of result helps pry apart two things that get tangled in casual talk: the pleasure of consuming sucrose and the motivational push produced by a sucrose-predicting cue. PIT, by design, tests cue power even when consumption is not happening.

PIT also hints at why people relapse after long abstinence. A cue can revive a whole action plan even when the drug itself is absent. The environment reawakens a learned policy: go there, do that, seek now. The pleasure of the drug is not required in that moment. The cue is sufficient to reanimate seeking.

Optogenetics allows this idea to be taken from "linked to" to "built by." Brittany Saunders and colleagues (2018) used phasic excitation of dopamine neurons paired with a neutral cue to create, in effect, a manufactured conditioned stimulus. A cue that had meant nothing gained the ability to drive later cue-locked behaviors. Even more revealing, dopamine neurons were not all interchangeable teachers. Subpopulations defined by their projection target could endow cues with different motivational properties. The same general cell type—dopamine neurons—could teach different lessons depending on where in the brain their axons delivered their message.

That projection detail is where the cue story becomes a story about negotiations among pathways rather than a single chemical flood. Dopamine neurons in the ventral tegmental area (VTA) project to several targets, including the nucleus accumbens (NAc) and parts of the prefrontal cortex (PFC). Those targets, in turn, project onward and interact with one another. A cue can recruit a "go" pathway that invigorates seeking, while control circuits attempt to dampen or reshape the response. Optogenetics makes it possible to perturb one projection while leaving another intact, temporarily tipping the negotiation.

The key trick is projection-specific manipulation. In practical terms, researchers express an opsin in a population of neurons (say, VTA dopamine neurons), then deliver light not at the cell bodies but at the axon terminals in a chosen target (say, the NAc). When the light hits terminals containing an excitatory opsin such as ChR2, it can drive release and downstream effects primarily in that projection. When the opsin is inhibitory, light can reduce neurotransmitter release or silence spiking in a time-locked way. The intervention is not "dopamine everywhere," but "this particular dopamine pathway, at this particular moment."

Inhibitory opsins are crucial for necessity tests. Halorhodopsin is a light-driven chloride pump that tends to hyperpolarize neurons under illumination, making them less likely to fire. Arch (archaerhodopsin) is a light-driven proton pump that also suppresses neural activity. These are not subtle, philosophical tools. In the right hands, they can shut down spiking with a flash and release it just as quickly. That temporal precision matters because cue responses unfold fast. The first fraction of a second after a cue appears is when attention shifts, action tendencies ramp up, and prediction systems update. If a circuit is briefly silenced during that window and the cue loses its grip, it suggests that circuit is not merely correlated with the cue's effect; it is participating in generating it.

Here is the kind of dissociation that makes the cue story feel less like a vague "reward system" and more like a set of levers. Silencing a pathway at cue onset can weaken cue-triggered seeking while leaving basic movement intact. An animal can still walk, groom, and explore. It can still consume rewards if they are placed in front of it. What changes is the cue's capacity to *pull* the animal into a specific, learned pattern of seeking. That tells you the manipulation is not simply sedating the animal or removing all pleasure; it is altering how opportunity signals get translated into action.

This is where earlier simplistic debates about dopamine can be put aside. If dopamine were only pleasure, then dampening dopamine should mainly flatten enjoyment, and cues should lose power only because the reward feels less good. But in many cue-driven paradigms the animal is not consuming anything during the critical moment. The cue is doing its work upstream of consumption. Necessity tests that target cue-time dopamine signaling therefore point to a role that is closer to "assigning motivational weight to a cue" than to "generating the hedonic sensation of reward."

The sign-tracking phenomenon offers an almost cinematic illustration of this. Watch a sign-tracker and a goal-tracker during the same lever-cue task. Both have learned. Yet the sign-tracker acts as if the cue itself is the thing to approach, even though the real payoff is elsewhere. In human terms it resembles what happens when the ritual becomes fused with the reward: the unlit cigarette in the hand, the clink of ice in a glass, the phone unlocked "just to check." These are not the rewards in any nutritional sense, yet they can be oddly gripping.

The concept of incentive salience helps make those observations less moralized and more mechanistic. Cues acquire incentive salience when they become attractive, when they elicit approach, and when they can energize actions that lead toward the reward. Saunders and Robinson (2012) argued for a selective role of dopamine in the expression of certain Pavlovian conditioned responses, especially those related to incentive motivation. In that framing, dopamine is not required for all learning about rewards, but it is pivotal for the particular kind of learning that gives cues their motivational magnetism.

Addiction-like behavior can then be seen as a specific outcome of circuit bargaining: the cue-valuation and invigoration pathways gain disproportionate influence over behavior, while other systems that encode long-term goals, competing needs, or contextual nuance lose their leverage at the crucial moment. Importantly, this can happen even in the presence of accurate knowledge. A person can know that using again is harmful, just as a goal-tracking rat "knows" that the food comes from the cup, not the lever. Knowledge does not always win an argument against a cue that has learned to shout.

The same logic also helps explain why opportunity cues can provoke a sudden surge of effort. In PIT, a cue that predicts sucrose can increase lever pressing even when the lever is temporarily unrewarded. That is not irrational; it is a policy that often works in nature. When a cue signals that reward is available, the world has changed: the opportunity cost of doing nothing rises. Dopamine transients in the nucleus accumbens that track this cue power make sense as part of a circuit that allocates effort when the environment is favorable. Addiction may exploit a system that evolved to seize time-limited opportunities.

Optogenetic experiments sharpen the point further by showing that the cue's properties depend on which dopamine pathway is engaged. Saunders et al. (2018) found that projection-defined dopamine subsystems could imbue cues with distinct motivational qualities. The phrase "projection-defined" sounds technical, but the intuition is simple: two teachers can both use chalk, yet one teaches calculus and the other teaches poetry, because they address different classrooms. Dopamine is the chalk. The axons determine the classroom.

This projection specificity is also a warning against treating dopamine as a single unitary signal. The same neurotransmitter can mean different things depending on where and when it is released, and which receptor types are present downstream. A cue-related dopamine transient in one striatal subregion can primarily invigorate seeking; in another, it may be more about learning or about movement preparation. When people speak of "dopamine spikes" in the abstract, the brain quietly asks: *in which circuit, during which computation?*

That question becomes emotionally salient when thinking about relapse. A cue that has been granted incentive salience is not just a reminder; it is a small motivational machine. Silencing the machine at the right instant can weaken cue-triggered seeking in an animal. That does not mean a human could be cured by turning off a single pathway. Human addiction involves stress, social context, pain, memory, and meaning, all braided together. Yet the circuit logic offers a humane clarification: compulsion can arise from learned, stimulus-bound control loops, not from a simple hunger for pleasure or a lack of willpower.

There is also an instructive historical echo here. Long before optogenetics, researchers knew that animals would work for stimulation of certain brain regions, and it was tempting to treat those regions as direct "pleasure centers." That language stuck in popular culture. What optogenetics added was the ability to separate the circuit components: not just where stimulation happens, but which cells, which projections, and which moments. When a cue can be given motivational force by pairing it with a phasic dopamine signal, and when that force can be selectively weakened by time-locked inhibition, it becomes harder to defend the idea of a single "pleasure dial." The brain begins to resemble a parliament: noisy, factional, sometimes deadlocked, sometimes captured by a loud minority.

Consider a familiar modern cue: the notification badge on a phone. The badge is not a reward in itself. It predicts a possible reward: social contact, useful information, a funny image, relief from uncertainty. For many people, that tiny red dot (or its equivalent) can provoke checking behavior that is out of proportion to the actual value of the message. It is not that the phone is chemically addictive in the narrow sense; it is that cues can be engineered to be persistent, frequent, and hard to ignore, and the brain's cue-valuation machinery is built to treat repeated predictors as worthy of attention and action. In the laboratory, a lever can become a magnet. In life, a sound can.

The optogenetic lesson is not that all cues are equally dangerous, or that dopamine is destiny. The lesson is that cue power is a property that can be assigned, amplified, and in some cases selectively interrupted. A cue that "steals the steering wheel" is not a mystical entity; it is a learned control signal backed by specific pathways. Those pathways can sometimes be teased apart: some contribute to learning that a cue predicts reward; others contribute to the cue becoming alluring and action-invigorating. That separation is precisely what makes addiction-like behavior intelligible as circuit negotiation rather than as a single craving molecule.

When a once-neutral cue starts to dominate behavior, it is tempting to say the animal is chasing pleasure. Often it is chasing *the chance* of pleasure, or more precisely, it is responding to a cue that has been granted motivational priority by dopamine-linked circuitry. The cue has become a proxy that can activate seeking even before any reward arrives, and sometimes even when the reward no longer arrives at all. Under those conditions, the most honest description is also the most unsettling: the behavior is being driven by learned opportunity signals that have acquired the authority to act first and ask questions later.

10.3 Wanting Is Not Running

A rat presses a lever more times than usual. A mouse sprints toward a port with unusual speed. A monkey works longer, tolerating more effort, to earn the same sip of juice. Those are the kinds of observations that made dopamine famous: turn dopamine up, and animals *do more* to get rewards. Turn it down, and they seem apathetic. From a distance, it is tempting to compress all of this into a single story about desire.

Up close, the story frays. "Doing more" can mean at least two very different things. It can mean that the animal assigns higher value to the outcome and is making a deliberate trade: more work is worth it. Or it can mean that the animal's motor system has been put into a more excitable state: movements start more readily, run faster, and repeat more often, even if the value calculation has not changed very much. Add a third possibility and confusion becomes almost guaranteed: a manipulation can also change *learning* so that the animal expects the reward more strongly, which then changes behavior later. The same outward behavior—extra lever presses—can be produced by an altered expectation, an altered motivational drive, or an altered ability to initiate and sustain movement.

This is why optogenetics is at its most persuasive when it is paired with careful recording and task design. If the goal is to claim "this circuit controls wanting," it is not enough to show that stimulating it increases pressing. You have to rule out a simpler explanation: that stimulation simply made the animal more active.

The confusion has a long history, and it is not limited to dopamine. Early brain stimulation studies often relied on measures such as how often an animal pressed a lever for stimulation. Those experiments were revolutionary, but they came with a built-in ambiguity. Lever pressing is not a pure readout of pleasure, or of motivation, or of learning. It is a small motor act repeated many times, and motor repetition is precisely the kind of thing that can be inflated by arousal or by a release of movement constraints. A similar ambiguity shadows modern work whenever the behavioral measure is "more of something," especially something that looks like vigorous movement.

Dopamine sits at a particularly tricky crossroads because of where it acts. Dopamine neurons project heavily to the striatum, a major input structure of the basal ganglia. The basal ganglia are not a "motor area" in the simple sense, but they are deeply involved in selecting actions, initiating them, and regulating their vigor. Their architecture offers a mechanism by which dopamine can bias action selection and invigoration without being equivalent to movement itself. Striatal circuits are often described in terms of a direct pathway that tends to facilitate selected actions and an indirect pathway that tends to suppress competing actions. Dopamine can shift the balance, making it easier for certain actions to win the competition for expression. That is not the same as moving a limb; it is closer to opening and closing gates that determine which movements get launched and with how much force.

The everyday intuition for this is familiar. Think of a morning when you are physically capable of going for a run but cannot initiate it. Then consider the same body after strong coffee, a sudden deadline, or a piece of exciting news. The muscles did not change dramatically in a minute. The gating did. "Wanting" and "running" are entangled partly because wanting often expresses itself through running, but the gate that allows running can be opened by many signals that have little to do with the pleasure of the reward.

The cleanest way to pull these threads apart is to pay attention to timing. If dopamine primarily controls movement execution, then altering dopamine while an animal is already moving should strongly affect the ongoing movement. If dopamine primarily gates initiation and energizes future movements, then its critical window should be *before* movement begins. Joaquim Alves da Silva, Fatuel Tecuapetla, and colleagues (2018) provided influential evidence on this point. They found that dopamine neuron activity before action initiation can gate the probability of initiating movement and influence future movement vigor; manipulations after initiation have little effect on ongoing movement (da Silva et al., 2018). The detail that matters here is not a technical one, but a conceptual one: dopamine can act like the starter motor rather than the steering wheel.

That changes how one interprets many "dopamine makes animals do more" results. If a burst or transient increase in dopamine arrives right before the decision point—right before the animal could begin a bout of pressing—it may increase the chance that a bout begins, and it may increase the pace once it starts. From the outside, that can resemble an increase in motivation or in reward value. It may *feel* like desire in a human. But in the circuitry, the dopamine signal may be working as an action gate and a vigor knob.

To make progress, experiments increasingly combine perturbation with recording. Recording can mean electrophysiology (listening to spikes), or calcium imaging (watching fluorescent signals that track activity in populations of neurons). The point is not that recording provides the truth while perturbation provides theater. The point is that each method constrains interpretation. If you stimulate dopamine neurons and see increased lever pressing, you still do not know whether you changed value, movement gating, or learning. But if you also record dopamine dynamics and downstream circuit activity while animals perform tasks designed to separate these components, the space of plausible explanations shrinks.

A key design move is to separate wanting from consumption. Consumption is when liking has its best chance to show itself—tongue movements, facial reactions, the gradual satisfaction of hunger. Wanting often shows itself earlier, in approach and in work. If a manipulation changes pressing without changing consumption patterns when the reward is delivered freely, that suggests a shift in motivation or action gating rather than a shift in pleasure. If a manipulation changes cue-evoked approach but leaves the sensory enjoyment of the reward largely intact, that again points toward incentive and action systems rather than hedonic ones.

That separation aligns with a central idea from Kent Berridge and Terry Robinson's work: dopamine's motivational "wanting" (incentive salience) can be dissociated from pleasure ("liking") and from the mechanics of movement execution. In everyday life, the dissociation is easiest to see when people pursue things that no longer feel good. Someone with a habit may report that the first few experiences were pleasurable and later ones were not, yet cues can still provoke a strong pull. Wanting can spike in the absence of liking. The nervous system is capable of treating "go get it" as a command that does not require "you will enjoy it" as a premise.

A concrete laboratory analog is the cue-triggered surge: a tone or lever that has become meaningful suddenly increases approach or pressing even when the reward is unavailable or when the animal is sated. The cue is not delivering pleasure; it is delivering a state change. It is a prompt to act. That is why cue-driven behavior is such a danger zone for interpretation. A manipulation that increases cue-triggered responding could be amplifying incentive salience—wanting—or it could be amplifying arousal and movement readiness, which also increase cue-triggered responding. Distinguishing these requires more than a single behavioral measure.

Computational accounts offer a useful language for this without turning it into mathematics. One influential idea is that dopamine levels relate to vigor by encoding average reward rate or opportunity cost: when the environment is rich with reward, moving slowly and hesitating is costly, so actions should be initiated quickly and pursued with energy. That framing aligns with evidence from Hamid and colleagues (2016) that dopamine tracks reward rate over minutes, and with Mohebi and colleagues (2019), who argued for dissociable dopamine dynamics for learning and motivation. Even without the equations, the intuition is clear. Dopamine is not only a scorekeeper for a just-delivered reward; it can also be a signal about whether it is worth getting off the couch.

This helps explain an otherwise puzzling observation: an animal can show reduced dopamine and still be capable of movement, yet appear "unmotivated." It also explains why dopamine manipulations can affect the *latency* to begin an action—how long the animal waits before it starts—more than the fine details of the movement itself. Latency is where motivation and movement gating overlap most tightly.

Here is a thought experiment that echoes the da Silva finding. Suppose a mouse is trained to run down a corridor to a port when a light turns on. If you stimulate dopamine neurons a fraction of a second before the light, you might see quicker starts and faster runs. If you stimulate halfway down the corridor, mid-stride, the effect might be much smaller. The same dopamine signal can appear powerful or weak depending on when it arrives. That is the signature of a gate.

The basal ganglia provide a plausible substrate for this timing sensitivity. Movement initiation is not a single moment but the resolution of a competition: hold still versus go; turn left versus right; approach versus retreat. Dopamine can tilt that competition, pushing the system toward launching an action. Once the action is underway, other circuits—motor cortex, brainstem pattern generators, spinal cord—carry the details. Under that view, dopamine is not the engine's pistons; it is closer to the ignition and the accelerator.

That analogy needs restraint, because too much "dopamine equals accelerator" talk risks becoming yet another single-story. Dopamine can also teach, as the millisecond-scale teacher. It can also assign motivational magnetism to cues. These roles are not identical, and part of the modern picture is that different dopamine dynamics and different projection-defined subsystems contribute in distinct ways. But the movement-gating role is an especially common source of misinterpretation, precisely because it can masquerade as wanting.

Recording helps catch that masquerade. If calcium imaging shows that a neural signal is tightly locked to movement onset across many conditions—rewarded, unrewarded, cued, uncued—it suggests a motor or action-initiation role. If, instead, the signal scales with expected value even when movement is held constant, it suggests a motivational or evaluative role. Good experiments often include task variants that force those variables apart. For example, animals can be trained to make the same movement for rewards of different sizes, or to choose between actions with different costs. If a signal tracks expected reward while movement remains similar, it is less likely to be mere movement. If it tracks movement vigor when reward is unchanged, it is less likely to be a pure value signal.

Optogenetics becomes more honest in this context too. Stimulation is not a magical probe that reveals "the function" of a neuron. It is a perturbation, and the behavioral output reflects the whole circuit's response to that perturbation. If stimulating dopamine neurons makes an animal press more, that might mean dopamine is a pleasure signal; it might also mean dopamine increases the probability of initiating a press sequence. Pair the same stimulation with high-speed video tracking and with recordings from downstream striatal neurons, and one can ask sharper questions: does the stimulation change the animal's posture, its general locomotion, its orienting responses? Does it change reaction time more than it changes choice? Does it change the structure of actions—longer bouts rather than isolated presses? Those details begin to reveal whether the intervention altered wanting, altered running, or altered the gate that links the two.

The practical stakes are not academic. Treatments for addiction, depression, and Parkinson's disease all brush up against dopamine systems, and each condition highlights a different facet of the wanting-running knot. In Parkinson's disease, dopamine loss in motor-related circuits produces profound difficulty initiating movement, often described by patients as if the body will not start, even when the desire to move is intact. In addiction, people can report strong wanting triggered by cues even when the pleasure has faded, and the wanting can be bound to rituals that are, at base, action sequences. In depression, the complaint is often not the inability to move but the inability to find anything worth initiating movement for—anhedonia and avolition interlaced. These are not identical, but all can involve a miscalibration of how easily the brain opens the gate to action and how strongly it tags opportunities as worth pursuing.

The cleanest conceptual payoff is a kind of humility about behavioral readouts. "Pressing more" is not a unitary sign. It is an outcome that can be produced by enhanced learning, intensified incentive salience, increased action initiation probability, greater vigor, reduced competing behaviors, or simple motor disinhibition. When optogenetics is paired with recording, and when tasks are designed to separate these possibilities, dopamine stops being a single dial and becomes what it has always been biologically: a set of signals that can teach, energize, and bias choice, each on its own timescale.

Wanting, then, is not a synonym for moving. It is a state that recruits movement by pushing on gates, by raising the gain on opportunities, by making certain actions feel urgent to initiate. Running is the visible consequence, but the deeper action is the moment a cue or expectation persuades the basal ganglia that now is the time to go.

10.4 When Repetition Hardens into Need

A first cigarette, for many people, is a lopsided bargain. The taste can be harsh; the dizziness is not always pleasant. The pull comes later, after the brain has done what brains do best: stitch together a set of reliable links between moments and outcomes. The walk to a familiar corner shop, the first breath of cold air outside, the click of a lighter, the small pause between inhalation and exhalation—each can become a kind of microscopic promise. Long after the pleasure has thinned, the promise can remain strangely loud.

Addiction is often described as "chasing a high," but that phrase can be misleading in two ways. First, it suggests a single target—pleasure—when the underlying machinery is distributed across circuits that teach, energize, and assign incentive salience to cues. Second, it suggests a fixed state, as if someone becomes addicted all at once. In the laboratory and in the clinic, a more accurate picture is gradual. Repetition does not merely strengthen a behavior; it can change the *kind* of control that behavior is under.

A widely used framing, sharpened over decades of work and articulated clearly by Barry Everitt and Trevor Robbins, treats addiction as a progression from goal-directed action to habits and then to compulsions. Goal-directed actions are sensitive to outcomes. If the outcome loses its value, the action declines. If the action stops producing the outcome, the action declines. Habits are less flexible. They can persist when the outcome is devalued, and they are more readily triggered by cues. Compulsions go further: behavior continues despite negative consequences, as if the cost-benefit calculation has been pushed to the margins.

No laboratory task captures the full human reality of addiction—its social context, its shame and secrecy, its entanglement with pain and trauma, its legal and economic pressures. But animal models can test specific causal claims: which patterns of activity are sufficient to reinforce behavior, which pathways are necessary for cue-triggered relapse-like responding, and how repeated exposure shifts the locus of control across circuits. Optogenetics earns its place here by tightening the logic. It can show not only that a circuit is active during relapse-like behavior, but that transiently altering that circuit at the right moment can alter the behavior.

The first bridge from reinforcement to addiction is straightforward: reinforcement strengthens action. If an animal receives a rewarding outcome after pressing a lever, it presses more. That could be food. It could be a drug. Or, in some optogenetic experiments, it can be a phasic pattern of dopamine neuron activity. But a strengthened action is not yet addiction. Many reinforced actions remain under goal-directed control: the animal presses when hungry, stops when sated, and adapts when contingencies change.

The crucial shift involves *repetition with cues.* Every reward episode is wrapped in a context: sights, smells, internal states, small rituals, the time of day. Over time, predictors can gain incentive salience and become capable of invigorating seeking on their own. The laboratory has an unromantic way of assessing this: remove the reward, present the cue, and see whether the animal still works. If it does, the cue has acquired power that is not reducible to immediate pleasure.

This cue power can become stronger with repeated drug exposure in a way that fits incentive-sensitization theory, associated especially with Berridge and Robinson: repeated drug experience can sensitize cue-triggered wanting, making relapse more likely even when hedonic impact is diminished. People often report this in plain language: "I don't even enjoy it like I used to, but I still want it." Incentive sensitization offers a mechanistic translation: the motivational system that assigns "wanting" to cues becomes hypersensitive, while liking does not necessarily increase and may even fall.

What does the brain do during that long arc? One influential idea is that control shifts from more ventral striatal circuits (including the nucleus accumbens) toward more dorsal striatal circuits, and from prefrontal control toward striatal, stimulus-bound control. The ventral striatum is often tied to flexible valuation and to integrating motivational signals. The dorsal striatum is strongly implicated in habits and in stimulus-response learning. This is not a simple handoff, as if one circuit clocks out and another clocks in. It is closer to a reweighting: the same cue increasingly drives behavior through circuitry that is faster, more automatic, and less sensitive to current goals.

The evidence for such shifts is not purely anatomical; it is temporal and functional. A striking illustration comes from studies of repeated cocaine self-administration showing that phasic dopamine signaling patterns across the striatum can shift over time, consistent with increasing recruitment of circuits associated with habitual control. The specifics vary by protocol, but the conceptual result is consistent: the dopamine system's most informative transients do not stay put. As experience accumulates, cue- and action-related dopamine signals can appear in striatal territories increasingly associated with habitual responding. The teaching signal and the invigoration signal, in other words, can migrate in their downstream impact.

That migration suggests a plausible mechanism for the lived experience of compulsion. Early on, drug taking can resemble a choice: an outcome is valued, and the person seeks it. Later, the behavior can feel like it is launched by triggers. A cue does not merely remind; it initiates. The system has learned a fast route from stimulus to action, and repeated dopamine-mediated plasticity may have helped pave that route.

Optogenetics can test parts of this story with unusual clarity by creating a simplified, controlled version of reinforcement and then asking whether repetition alone can generate relapse-like features. A key demonstration comes from work by Pascoli and colleagues (2015): optogenetic dopamine neuron self-stimulation can be sufficient to produce addiction-like progression features, including cue-induced relapse after abstinence, alongside synaptic potentiation onto D1 accumbens neurons. The phrase "self-stimulation" can sound like a parlor trick until one remembers what is being claimed. If animals will work for phasic dopamine activation, and if repeated self-stimulation can later produce cue-triggered seeking after a period of enforced abstinence, then dopamine bursts are not merely pleasant events. They can function as the kind of reinforcement that builds a durable cue-action loop.

The synaptic detail matters because it anchors behavior to a plausible cellular mechanism. D1 receptors are one of the main dopamine receptor types in striatal circuits, often associated with the direct pathway and with action facilitation. "Synaptic potentiation" means that particular synapses have become stronger: the same input now has more influence on the neuron. If repeated dopamine self-stimulation strengthens excitatory inputs onto D1-expressing neurons in the nucleus accumbens, that provides a biological substrate for why cue-related inputs might become more effective at driving downstream action. The behavior is not simply a mood; it is a rewired weighting of inputs.

At the same time, an honest account has to draw boundaries around what such experiments prove. When an animal presses a lever to activate its dopamine neurons, the experiment has isolated one component of drug action and magnified it. Many drugs do more than increase dopamine transients. They have pharmacological effects on multiple neurotransmitter systems, they alter sleep and stress hormones, they change learning through aversive withdrawal states, and they interact with social environment. Optogenetic dopamine self-stimulation therefore cannot be equated with "modeling cocaine" or "modeling heroin" in full. What it can do is test a sharper claim: repeated phasic activation of dopamine neurons is sufficient to build persistent seeking and cue-triggered relapse-like responding under certain conditions.

The same kind of boundary setting applies to the language of compulsion. In humans, compulsion is a clinical and moral weight-bearing word. It implies a loss of agency, a narrowing of choice, a kind of suffering. In animals, researchers operationalize compulsion-like behavior using specific criteria: persistence despite punishment, persistence when the outcome is devalued, high rates of seeking triggered by cues after abstinence, resistance to extinction. These are meaningful, but they are not identical to the human experience. A rat that continues seeking despite a mild foot shock is not "choosing self-destruction" in the way a person might. It is displaying a measurable rigidity of behavior under cost.

This is where the experimental logic itself becomes part of the story. Addiction is not one phenomenon but an umbrella under which several processes gather: learning of cue-reward associations, attribution of incentive salience to cues, changes in action vigor and initiation thresholds, habit formation, stress-linked relapse, and the aversive pull of withdrawal. Each can be tested with distinct paradigms, and optogenetics can be deployed with distinct kinds of claims in mind. A manipulation that changes cue-triggered seeking does not automatically change consumption. A manipulation that increases vigor does not automatically increase pleasure. A manipulation that alters learning does not necessarily alter the immediate urge. Without that care, it is easy to oversell, to declare a "circuit of addiction" when the data support something more precise and more useful: a circuit component that participates in a particular stage of a particular task.

A helpful way to picture the progression from goal-directed to habit-like control is to think in terms of where flexibility lives. Goal-directed behavior leans heavily on the ability to represent outcomes, to compare them, and to incorporate current needs. Prefrontal and related cortical networks are well suited to that, with their capacity for context and planning. Habit-like behavior leans on efficient stimulus-response mappings: if cue, then action. The striatum is well suited to building such mappings, especially when dopamine signals repeatedly stamp certain actions and cues as significant. Over time, the "if cue, then action" route can become faster than deliberation, and in moments of stress or distraction, faster often wins.

Stress deserves special mention because it is one of the most potent accelerants of relapse in people, and it can be modeled in animals in limited but informative ways. Stress changes dopamine dynamics, shifts attention toward immediate opportunities, and can impair prefrontal control. Under stress, a cue that once could be resisted can again feel commanding. This does not require that stress makes the drug more pleasurable; it may make the cue-action loop easier to launch and harder to brake. In that sense, stress may not add value so much as subtract control.

The darker implication is that "recovery" is not always erasure. Extinction training, whether in a clinic or in a cage, often builds new learning that competes with old learning rather than deleting it. A cue can be rendered less powerful in a safe context and then regain power in a different context, a phenomenon known as renewal. The brain, being cautious, keeps old threat and reward associations available. That caution is adaptive in the wild and cruel in addiction.

Yet the same mechanistic picture also contains a kind of hope, or at least a reframing. If addiction is a circuit-level trap rather than a moral failure, then relapse is not proof of weakness; it is evidence that certain learned loops remain potentiated and can be reactivated under particular conditions. That does not excuse harmful actions, but it clarifies why exhortation alone so often fails. You cannot argue a potentiated synapse out of existence.

A final, sobering point concerns measurement. Many of the most compelling demonstrations of circuit shifts and relapse-like behavior come from carefully controlled settings: a single drug, a single lever, a small set of cues. Human lives are not controlled settings. People encounter layered cues, competing rewards, complex stressors, and social consequences that can themselves become cues. Translating from mice to medicine therefore requires both ambition and restraint: ambition to use causal tools to identify leverage points in circuits, and restraint to avoid declaring that a single circuit manipulation has "explained addiction."

Repetition changes brains. That statement is banal until one watches, in data and in behavior, how repetition narrows the funnel from cue to action. At first the action is a choice pursued for an outcome. Later the outcome becomes less central than the cue, and the cue becomes a compressed trigger that can launch seeking with surprising force. Somewhere along the way, the brain's teacher signals and gate signals stop serving flexible goals and start serving a loop that runs by default.

The most haunting part is that the loop can be built from ordinary materials: prediction, reinforcement, incentive salience, action gating. None of these are pathological on their own. They are the tools of learning and survival. Addiction, in this light, is not an alien process but a familiar one pushed into an extreme regime, where the lessons are too strong, the cues too potent, and the gates too easily opened.

Chapter 11

Mood Circuits and Second Chances: From Depression to Restored Sight

A pulse of light can do more than make a mouse freeze or run; in the right circuit, it can hint at why despair lifts, why timing matters as much as location, and why the eye may be the first place optogenetics truly helps patients. By the end, the reader sees both the seduction and the sobriety of the field: a technology that promises not magic control of the mind, but sharper, more testable ideas about repair, responsibility, and hope.

11.1 When Sadness Becomes a Wiring Problem

For much of the modern era of psychiatry, depression has been narrated as a shortage story. Not enough serotonin. Not enough norepinephrine. Perhaps not enough dopamine. The story gained its persuasive power from a striking historical accident: some of the earliest antidepressants were discovered while doctors were looking for something else.

In the 1950s, the tuberculosis drug iproniazid was noticed to lift patients' spirits. Around the same time, imipramine, synthesized as a possible antipsychotic, turned out to help some patients who were sunk in melancholy rather than agitation. These drugs did not arrive with a tidy theory attached; they arrived with a clinical surprise. Only later did researchers work backwards to biochemical mechanisms, and only later still did those mechanisms harden into public-facing explanations. If medicines that increase monoamine signalling can relieve symptoms, then perhaps depression is a monoamine deficiency. The logic is understandable, but it is also a classic scientific temptation: confusing a useful lever with the whole machine.

That temptation has consequences. A shortage story makes depression sound like a single thermostat set too low, waiting for the right chemical adjustment. It encourages a picture of mood as a smoothly changing dial, governed by one or two knobs. It offers comfort by suggesting that suffering has a straightforward biological counterpart. It also flattens the experience of depression into something that resembles fatigue or vitamin deficiency: unpleasant, yes, but conceptually simple.

Clinical reality never quite cooperated. Antidepressants that alter neurotransmitter levels within hours often take weeks to help, and sometimes do not. Many patients describe not a uniform sadness but a suite of changes: the world loses its pull; sleep and appetite distort; thought loops tighten; guilt becomes sticky; the future feels not merely uncertain but unavailable. Depression can numb rather than sadden. It can come with agitation rather than slowing. Some people cannot concentrate; others cannot stop ruminating. A single shortage mechanism struggles to explain why such different inner weather patterns share a diagnostic name, or why treatments with very different immediate actions can converge on relief.

A circuit view does not deny chemistry; it relocates chemistry into a larger causal map. Neurotransmitters become ingredients in the way networks operate, not the entire recipe. Mood, on this view, is not a substance in the brain but an *activity pattern* distributed across interacting systems: prefrontal regions involved in planning and regulation; limbic structures that tag experiences with threat or safety; and reward-related pathways that help decide what is worth effort. Depression begins to look less like an empty reservoir and more like a set of maladaptive dynamics: loops that get stuck, signals that amplify, and networks that fail to coordinate.

This change in framing was not merely philosophical. It was driven by tools that could finally ask causal questions with the kind of precision that earlier methods could only approximate. As we saw with optogenetics, turning selected neurons on and off with light is not just a technical trick; it is a way of testing claims about what a circuit *does*. If a brain area is quieter in a depressed state, does increasing its activity help, or is the quietness an effect rather than a cause? If a pathway is more active under stress, is that activity protective, or is it part of the pathology? Correlations alone cannot tell you. Optogenetics can, at least in animals, start to pry those questions apart.

A concrete example comes from work by Covington III, Lobo and colleagues (2010) using a chronic social defeat stress paradigm in mice. The details of the paradigm matter. A mouse is repeatedly exposed to an aggressive peer, a controlled experience that produces enduring changes in social behaviour in a subset of animals: some become socially avoidant and show behaviours interpreted as despair-like or anhedonia-like, while others remain relatively resilient. These are not human symptoms, and no responsible scientist claims otherwise. But the paradigm creates something valuable: a stable behavioural shift tied to stress exposure, in an animal whose brain circuits can be manipulated and measured.

Covington and colleagues focused on the medial prefrontal cortex (mPFC), a region often discussed in relation to executive control, appraisal, and the ability to regulate emotion. Human imaging studies had repeatedly found altered prefrontal activity in depression, but imaging cannot tell you what is cause and what is consequence. In the stress-susceptible mice, the researchers used optogenetics to drive mPFC neurons with a pattern designed to resemble burst firing. The striking result was that this stimulation reduced certain depression-like behaviours: mice became less avoidant and performed differently on assays commonly used to index behavioural despair. The effect was not easily explained away as mere hyperactivity or general arousal. The intervention looked, within the constraints of the model, like an antidepressant-like push delivered not by a drug bath but by patterned activity in a specific circuit node.

If that were the whole story, it would be seductively simple: depression as prefrontal underactivity, cured by turning the prefrontal cortex up. But optogenetics immediately complicated that tidy picture, because it allowed researchers to ask a sharper question: *which* prefrontal outputs? The prefrontal cortex is not a single switchboard. It is a hub with many downstream partners, and the same patch of cortex can send fibres to areas that pull behaviour in opposing directions.

Warden, Selimbeyoglu and colleagues (2012) offered an early, elegant demonstration of this pathway specificity. Rather than treating the mPFC as a monolith, they traced and manipulated a projection from mPFC to brainstem targets involved in responding to challenges. By stimulating or silencing a defined projection, they could bias how animals responded under behavioural demand. The conceptual shift is subtle but profound: instead of "prefrontal cortex regulates mood," the claim becomes "this prefrontal output channel participates in mobilizing coping responses under this kind of challenge." That is a claim a circuit can support or refute.

The same logic applies to reward circuitry. Dopamine neurons in the ventral tegmental area (VTA) are often described in popular culture as a pleasure system. In depression, where motivation and anticipatory pleasure often collapse, dopamine is an obvious suspect. Yet dopamine neurons are not a uniform fountain of pleasure; they are diverse, they project to different targets, and their firing patterns carry distinct messages. Chaudhury, Walsh and colleagues (2013) used optogenetics to drive VTA dopamine neurons in a phasic pattern, meant to mimic the brief bursts that occur naturally in response to salient events. In a social defeat context, phasic stimulation could promote susceptibility-like outcomes. Crucially, projection specificity mattered: VTA neurons projecting to the nucleus accumbens (NAc), a key node in motivation and action selection, were implicated in promoting these stress-related behavioural changes, whereas other projections behaved differently.

Put these findings side by side and the older shortage narrative begins to crumble. "More dopamine" is not automatically antidepressant. "More prefrontal activity" is not automatically beneficial. What counts is how activity is routed through a network, and how that routing interacts with the animal's current state.

A spare circuit sketch can hold this idea without drowning in anatomy:

The picture to keep in mind is not that any one arrow is "the depression pathway." Rather, mood emerges from the coordination (and miscoordination) among several loops. A prefrontal region that helps an animal persist through challenge must talk to brainstem and midbrain structures that implement arousal and action. Reward circuits must integrate context and expectation, deciding whether effort is worth it. Threat circuits must decide whether a social encounter is danger or opportunity. Depression can involve any combination of disruptions: a prefrontal system that fails to dampen threat; a reward system that assigns too little value to future outcomes; a stress system that tags neutral experiences as costly; or a network that cannot shift states efficiently.

This network framing also helps make sense of ketamine's unusual antidepressant profile. Ketamine can produce rapid relief in some patients, sometimes within hours, which is hard to reconcile with a simple story of slowly replenishing neurotransmitters. Part of the emerging account involves glutamate and synaptic plasticity, but circuit dynamics enter sharply through work on the lateral habenula (LHb), a small structure sometimes described as a hub for negative motivational value. Yang, Cui and colleagues (2018) reported that bursting activity in LHb was linked to depression-like behaviours in rodent models, and that ketamine blocked such bursting, aligning with rapid behavioural improvements. In this line of work, the relevant unit is not merely which chemical is present, but whether a node is firing in a pattern that can effectively hijack downstream systems. "Bursting" is a kind of punctuation in neural language: a way of making sure a message is heard.

Once you start thinking in these terms, some familiar puzzles look different. Consider the long lag between antidepressant drug ingestion and clinical improvement. A circuit account suggests that acute chemical changes may set the stage for slower processes: synaptic remodeling, shifts in connectivity strength, and relearning of behavioural strategies. A person who has been depressed for months has not only had altered neurotransmission; they have lived in a different statistical world. They have practised withdrawal. They have encountered fewer rewards. Their social network may have thinned. The brain is a prediction machine, and depression can be seen as a set of grim predictions repeatedly confirmed by behaviour and circumstance. Changing the predictions might require more than pushing neurotransmitters upward; it may require reconfiguring circuit responses so that new experiences can be registered as new.

Optogenetic experiments sharpen a second, equally important point: causation is not a magic wand. When a scientist drives a set of neurons and observes a change in behaviour, it is tempting to speak as if those neurons *caused* an emotion. But "emotion" is not something a mouse can report, and behaviours in laboratory assays are at best proxies. A mouse that stops avoiding another mouse is not necessarily "happier." It might be less anxious, more impulsive, less able to remember the social defeat, or simply more prone to exploration. Careful studies try to rule out these confounds by measuring locomotion, anxiety-like behaviour, memory, and sensory function. Still, the broader caution remains: optogenetics offers strong causal leverage over neural activity, but it does not automatically translate that leverage into a perfect explanation of inner life.

There is also a subtler interpretive trap. Optogenetics can make a circuit look like a simple on/off device because the experimental manipulation is crisp: light on, neuron spiking; light off, neuron quiet. Yet the brain's own control signals are rarely so binary. Stress hormones, neuromodulators, sleep pressure, hunger, and social context all alter excitability and synaptic gain. A stimulation that rescues behaviour in one state may do little in another. A "depression circuit" might be better described as a network configuration that becomes too stable, like a song stuck on repeat. In that sense, depression is not located in one place; it is distributed across a set of interactions that can hold the system in a narrowed repertoire.

That view also helps explain why the same anatomical region can be associated with opposite effects across different studies. Lüscher and Pollak (2019) emphasized how findings in depression-related circuitry depend on the specific stress model and behavioural assays used. Social defeat, chronic mild stress, learned helplessness, early-life stress: each produces a different constellation of changes, and each interacts differently with individual variability. In human terms, these are not interchangeable paths into depression; they resemble different biographies. It would be surprising if the underlying circuit signatures were identical.

All of this can sound abstract until it is connected back to lived experience. Think about the ordinary act of getting out of bed when you have something you want to do. That act recruits a coalition: a system that represents goals, a system that estimates effort cost, a system that predicts reward, a system that manages threat and uncertainty, and a body that must be energized to move. Depression can be experienced as the collapse of that coalition. The goal may remain, but it no longer recruits momentum. The estimate of effort may swell. The predicted reward may shrink to near zero. Threat signals may rise: the day looks full of hazards, including the hazard of failure. In a circuit framing, those are not separate symptoms glued together by diagnostic criteria; they are the psychological surface of a network whose internal negotiation has tilted toward withdrawal and pessimism.

The circuit view also forces humility about intervention. If depression is a network phenomenon, then there may be multiple effective entry points: medication that shifts neuromodulatory tone, psychotherapy that changes the informational diet of the network and teaches new control strategies, sleep interventions that reset oscillatory regimes, social reconnection that restores reward and safety signals. Optogenetics does not replace these; it clarifies what it would mean to have a truly targeted biological treatment. A targeted treatment would not merely "increase serotonin" but would push a particular pathway into a healthier regime at a time when the rest of the system can respond. It would respect that symptoms are not just chemical outputs, but behaviours and thoughts that also feed back, reshaping the circuitry through learning.

When sadness becomes persistent and disabling, it is not simply because a molecule ran low. It is because a brain, doing what brains evolved to do, has settled into a pattern that prioritizes retreat, conserves effort, and expects little from the future. Optogenetics, by letting researchers perturb defined pathways and observe the consequences, has made that pattern visible in a new way: not as a moral weakness or a vague imbalance, but as dynamics in a living network that can, under some conditions, be pushed into a different mode.

11.2 The Same Neurons, Different Songs

A familiar fantasy about the brain goes like this: find the right cells, flip them on, and you get the thought or feeling you want. It is a fantasy that optogenetics seems designed to encourage, because the tool itself looks so decisive. Light on; spikes appear. Light off; spikes fade. The temptation is to imagine that "turning on the fear neurons" or "activating the antidepressant pathway" is like pressing a labelled button.

Brains do not work that way. Not because they are mystical, but because timing is part of the message. Neurons speak in electrical pulses, and the meaning of a pulse depends on when it arrives, how it is grouped with other pulses, and what the surrounding network is already doing. The same cells can participate in reassurance or panic, in approach or avoidance, in vigour or collapse, depending on the cadence of their firing and the state of the circuit receiving it.

A useful historical foothold comes from the first time anyone listened to the brain's rhythms from outside the skull. In 1929, the psychiatrist Hans Berger reported recordings of what would later be called the alpha rhythm, a roughly 10 Hz oscillation detectable with electrodes on the scalp. The drama of that moment was not that he had "found" thoughts, but that he had found organized timing: a collective beat that waxed and waned with eyes open or closed. The brain, it turned out, does not merely become more or less active; it changes *mode*. That early observation has echoed through a century of neuroscience. Whether one records from the scalp, from a single neuron, or from a genetically defined cell population, the key is often not the average activity level but the patterning: synchrony, bursts, pauses, and their alignment across regions.

Optogenetics forces this lesson in a particularly stark way. When researchers stimulate neurons with light, they must choose a pattern: a steady train of pulses, brief bursts separated by pauses, or something that locks to an oscillation in the ongoing activity. Those choices are not cosmetic. They can turn the same group of neurons into a different signal.

Consider dopamine neurons in the ventral tegmental area. In casual conversation dopamine is treated as a synonym for pleasure, but the system is more like a teaching signal and a mobilization signal rolled together. Dopamine neurons can fire in a slow, steady manner or in brief, intense bursts. Those bursts are often called *phasic* firing: rapid packets of spikes that stand out against baseline activity. Phasic firing is not simply "more dopamine." It is a different kind of event, arriving at a particular time and often tied to cues that predict reward, surprise, or motivational relevance.

Chaudhury, Walsh and colleagues (2013) used optogenetics to drive VTA dopamine neurons in phasic patterns within a social defeat stress paradigm. The point was not to show that dopamine "causes depression," which would be far too crude, but to test whether a specific temporal signature of dopamine firing could bias vulnerability under stress. In that context, phasic stimulation promoted susceptibility-like outcomes, particularly through VTA neurons projecting to the nucleus accumbens. A reader might reasonably ask: how can a burst pattern that in other settings is associated with reward and learning promote a depression-related phenotype? The answer begins with context. Under repeated social defeat, the environment is rich in salient, punishing information. A phasic teaching signal, delivered at the wrong times or routed through the wrong projection, may strengthen the learning of avoidance and defeat rather than the learning of approach. The same "burst" can become a stamp of value, a stamp of threat, or a stamp of inevitability, depending on what the brain has been sampling from the world.

Prefrontal cortex provides another instructive case. Broadly speaking, prefrontal regions are involved in selecting actions, reappraising meaning, and regulating emotional responses. Yet "prefrontal activation" is not a single intervention. The prefrontal cortex contains multiple cell types, multiple microcircuits, and multiple output routes. Optogenetics revealed a further twist: downstream target can flip valence. Literature discussed alongside infralimbic prefrontal cortex (IL-PFC) work on rapid antidepressant-like actions (Fuchikami, Thomas and colleagues, 2015) highlighted that stimulating IL-PFC can reproduce ketamine-like effects in rodents, but that the behavioural sign of "antidepressant-like" is not guaranteed by the cortical location alone. The output pathway engaged can push the network toward relief or toward worsening. A prefrontal message routed to one downstream region may recruit coping, while that same prefrontal node routed elsewhere may intensify negative bias or rumination-like persistence in rodent assays. The anatomy gives you addresses; timing and routing determine what gets said.

The lateral habenula (LHb) offers perhaps the clearest illustration of "same neurons, different songs," because the relevant patterns are vivid and, in a way, intuitive. The LHb sits at a strategic junction, influencing monoamine systems and integrating signals about disappointment, punishment, and negative outcomes. Yang, Cui and colleagues (2018) proposed that a particular pattern in LHb—bursting—is a circuit signature linked to depression-like behaviours in rodents, and that ketamine's rapid antidepressant effects involve blocking this bursting. Bursting is not merely higher activity. It is a special punctuation mark: clusters of spikes fired in quick succession, separated by quieter periods. That punctuation can drive downstream targets in a way that a steady trickle cannot. A burst can force a synapse to release more transmitter, synchronize downstream neurons, and alter plasticity rules. In other words, it can be a way of insisting that a message be heard.

A striking implication follows. If depression-like behaviour in some models aligns with LHb bursting, and relief aligns with preventing that burst mode, then the relevant therapeutic target is not "the habenula" in the abstract. It is the *state* the habenula enters: tonic firing versus bursting. That is a vastly more precise claim than anything that could be made with a broad chemical narrative. It also makes clear why some interventions are rapid. If a circuit node is stuck in a pathological rhythm that can seize control of downstream networks, nudging it out of that rhythm may produce immediate behavioural changes, long before the brain has had time to regrow spines or rewrite gene expression. Those slower changes may still matter for durability, but the first domino is dynamical.

Timing and rhythm have another practical consequence: they determine what an optogenetic experiment is actually testing. A headline might say "activating these neurons produces resilience," but the core optogenetics principle is that "what you stimulate" includes cell type, projection, subcellular compartment, and temporal pattern; without those specifications, claims about "the same neurons" are often misleading (a point emphasized in translational discussions of precision and interpretive limits). If light is delivered to cell bodies in one experiment and to axon terminals in another, the circuit perturbation is not the same. If one lab uses sustained stimulation and another uses burst-like patterns, the nervous system may treat those as different messages, not different magnitudes. If stimulation occurs during a behavioural challenge in one case and during rest in another, the brain state is different, and state changes what inputs are gated and what plasticity rules are active.

That last point can be understood without technical machinery. Think about the human experience of being spoken to while you are wide awake and engaged versus being spoken to while half-asleep or overwhelmed. The words may be identical; what they do to you is not. The brain has analogous gates, built from neuromodulators, inhibition, and network synchrony. "Circuit inertia and state," as we saw earlier, can make an intervention either catch and propagate or vanish into noise. Optogenetics, precisely because it is such a clean perturbation, can reveal how strong those gates are.

This is where closed-loop approaches become more than a clever engineering flourish. In an open-loop experiment, stimulation is delivered according to the experimenter's schedule: every second, or during a predefined task epoch, regardless of what the brain is currently doing. In a closed-loop or activity-guided experiment, stimulation is timed to an endogenous event: a particular oscillatory phase, a burst in a target region, or a behavioural marker that correlates with a state transition. Research programs emphasizing closed-loop designs have highlighted state dependence directly: timing stimulation to endogenous rhythms or brain states can change outcomes compared with open-loop stimulation (S13). In mood-related circuitry, this is not a minor refinement. Depression is often described by sufferers as a trap: an inability to shift gears, to break out of a loop of thought or lack of motivation. If the pathology is, in part, a failure of flexible state transitions, then an intervention that ignores state risks pushing on a door that is currently bolted from the inside.

These considerations also sharpen the meaning of causality in circuit experiments. There is a plain-language distinction that helps: *sufficiency* versus *necessity*. If activating a pathway produces a behavioural shift, the pathway may be sufficient to bias the system under those conditions. If silencing a pathway prevents the shift, the pathway may be necessary, again under those conditions. Neither claim automatically tells you whether the pathway is the "real cause" in an everyday sense, because brains are redundant and adaptive. A pathway can be sufficient without being necessary, and necessary without being sufficient, depending on what other routes can compensate. Timing complicates this further. A pathway might be necessary only at a specific moment: for initiating a coping response but not for maintaining it, for encoding a negative memory but not for retrieving it, for shifting out of a burst mode but not for staying out.

This is why oversimplified interpretations of optogenetic mood experiments can be so misleading. A brief stimulation can function like an instruction injected into a network, but the network may treat that instruction in ways that depend on its current commitments. Under chronic stress, a reward cue may be coded as a reminder of failure; under safety, it may be coded as promise. Under high arousal, a prefrontal input may recruit control; under fatigue, it may fail to propagate. An optogenetic pulse train is not a generic "activation"; it is a language with grammar.

That grammar, in turn, suggests why psychiatric treatments have been so hard to make precise. A deep brain stimulation electrode, for example, can deliver regular electrical pulses, but it cannot by itself specify a cell type, a projection, or a microcircuit, and it rarely adapts in real time to the patient's changing brain state. Drugs can reach cell types and synapses broadly but cannot readily deliver millisecond-specific patterns to a chosen pathway. Optogenetics, in animals, can do both—but it also reveals just how much specificity is required before a claim becomes stable across contexts. The more precisely one asks the question, the more the answer depends on the exact form of the question.

There is a humanly resonant way to hold onto all this without drowning in technicalities. Depression is often experienced as a narrowing: the same thoughts return; the same predictions win; the same small set of behaviours repeats. The brain, at its best, is musically versatile. It can hold a steady rhythm during routine action, punctuate with bursts during learning, synchronize across regions during attention, and loosen synchronization during creative exploration. A depressed brain may not be "quiet" or "chemically depleted." It may be playing the wrong rhythms too often, or failing to switch rhythms when the world changes. The tragedy is not a missing instrument; it is an inflexible score.

Optogenetics does not offer a simple cure for that tragedy, and it does not provide a direct window into subjective suffering. What it does offer is an uncompromising lesson in neural semantics: meaning is carried not only by *who* fires, but by *when*, in *what pattern*, and against *which background state*. When the same neurons can push behaviour toward vulnerability under one timing regime and toward resilience under another, the old urge to hunt for a single biological essence of depression gives way to a more realistic, and more humane, picture: mood as a property of coordinated dynamics that can, under the right conditions, be retuned.

11.3 Why the Eye Went First

If optogenetics is a way of speaking to neurons in the language of light, then the eye has always been the most natural place to try a first conversation with patients. Not because vision is simple—it is one of the brain's most elaborate accomplishments—but because the front end of the visual system sits out in the open, already designed to receive photons, and already connected to the rest of the brain through a well-traced cable of axons.

The basic practical problem of optogenetic medicine is easy to state and hard to solve: you need to deliver genes to the right cells, then deliver light to those cells safely and in a pattern the brain can use. In the deep brain, each part of that sentence is a formidable obstacle. Gene delivery requires reaching tissue protected by bone, membranes, and a vigilant immune system; light delivery requires implanted hardware because light does not travel far through scattering brain tissue; and measuring benefit is often indirect, because a person's mood, memory, or compulsions are not as straightforward to quantify as a reflex.

The retina softens all of those obstacles at once. It is optically accessible. It supports direct light delivery. It sits in an organ where gene therapy workflows were already becoming established. And it offers unusually clear readouts, from basic measures such as whether a person can detect a flash of light to more functional measures like locating an object on a table. That combination made retinal optogenetics a plausible early route from elegant laboratory demonstrations to the first careful steps in humans.

The retina is not merely a sheet of passive sensors. It is a layered circuit that performs substantial computation before signals ever reach the brain. Photoreceptors (rods and cones) capture light; downstream interneurons refine the signal; and retinal ganglion cells send the output to the brain via the optic nerve. In many blinding diseases, the tragedy begins with the photoreceptors. In retinitis pigmentosa, for example, photoreceptors gradually die, often leaving downstream retinal neurons—including ganglion cells—still present for some time. That creates a tantalizing opportunity: if photoreceptors are gone, could surviving retinal cells be made light-sensitive, so that the retina can again produce a meaningful output?

That is the conceptual leap of retinal optogenetic therapy. Instead of repairing or replacing the damaged photoreceptors directly, the strategy can bypass them: deliver an opsin gene to surviving retinal neurons so that those neurons respond to light. The opsin is a light-gated protein; once expressed, it can turn photons into electrical changes in the cell, leading to spikes that travel toward the brain. The promise is not to restore normal vision, but to restore a channel of information.

Why should this be easier than doing optogenetics for mood or memory? Start with the simplest constraint: light must get to the opsin. The skull is an excellent helmet and an inconvenient optical barrier. Brain tissue is also strongly scattering; blue light, in particular, does not penetrate far. So deep brain optogenetics tends to require implanted optical fibres or LEDs placed close to target tissue. The retina, by contrast, is already exposed to light through the pupil. An ophthalmologist can view it directly. Delivering light is not the hard part—the hard part is delivering the *right* light.

That statement contains the second advantage: the retina's input is naturally optical, so there is an obvious way to define and measure what an intervention should do. You can present a visual stimulus and observe whether the retina responds, whether the brain responds, and whether the person's behaviour improves. The intervention sits at the entrance to a sensory pipeline where stimulus and output are both tangible. This does not make the science trivial, but it makes the feedback loop clearer than in many psychiatric applications, where the "stimulus" might be a stressful day and the "output" might be a shift in self-reported hopelessness.

Gene delivery provides the third advantage. Viral vectors such as AAV (adeno-associated virus) have become a central tool of gene therapy because they can deliver genetic cargo to cells with relatively durable expression. The eye has been a major proving ground for this broader gene-therapy ecosystem. It is a small, enclosed compartment; clinicians can deliver vector locally; and the organ can be monitored closely over time. That infrastructure does not eliminate risk, but it lowers the barrier to an initial clinical attempt compared with deep brain delivery, where one must weigh neurosurgical risks and the uncertainties of long-term expression in delicate tissue.

These advantages help explain why, by 2021, a clinical proof-of-concept became possible. A Nature Medicine study by Sahel, Boulanger-Scemama and colleagues (2021) reported partial recovery of visual function in a blind patient after optogenetic therapy. The intervention combined AAV-delivered opsin expression in retinal ganglion cells with engineered goggles that delivered a tailored light stimulus. The case is often described as a milestone, and it is—but it is also a lesson in how much "translation" really means. The result was not a light switch that flipped vision back on. It was a new interface between the world and the brain.

The need for goggles is central to that lesson. In ordinary vision, the retina receives a continuous, rich image. Photoreceptors convert that image into signals that the retinal circuit transforms into a language the brain has spent a lifetime learning to interpret. In optogenetic vision restoration, the retinal cells being targeted were not designed to serve as photoreceptors. They are downstream elements with different properties: different sensitivity, different temporal response, different wiring. The opsins they express are not identical to native phototransduction machinery. As a result, the raw visual world must often be preprocessed and translated into a pattern of light that is compatible with the engineered retina.

Goggles or headsets do that translation. They capture the scene with a camera, process it in real time, and project a simplified, high-contrast representation onto the retina using LEDs or similar light sources. This is not a cosmetic add-on; current approaches often require external devices to transform visual scenes into an encoding compatible with the opsin and to manage intensity and spectral constraints (S9, S17). In other words, optogenetic therapy here is best understood as a co-designed system: biology plus device.

Why can't the world's ordinary light suffice? Because opsins have their own sensitivities, and safety imposes constraints. The retina is precious tissue; intense light can cause damage. Wavelength matters too. Many commonly used opsins in laboratory neuroscience respond strongly to blue light, but blue light carries higher photon energy and poses particular concerns for safe retinal illumination at the intensities that might be required. For clinical translation, red- or amber-shifted opsins—proteins tuned to longer wavelengths—offer an advantage: they can be activated with light that is generally considered more compatible with safety constraints for retinal exposure. Preclinical primate work has evaluated clinically plausible illumination and pattern discrimination using such red-shifted choices (S10). The point is not that red light is inherently harmless, but that the engineering space becomes more realistic when one can work at wavelengths and intensities that fit within conservative safety margins.

The device also helps with sensitivity. A person with retinitis pigmentosa may have little or no functional photoreceptor input. If an opsin is expressed in ganglion cells, those cells may require light brighter than ordinary indoor illumination to respond reliably. Goggles can provide controlled, intensified stimuli, as well as adapt the image so that edges and motion cues—the kinds of features the brain can make use of quickly—stand out.

A second lesson from the human case is time. The therapy did not yield its benefits the moment the vector was injected. There is a biological wait: time for the opsin to be expressed in the target cells. And then there is a behavioural wait: time for the person to learn what the new signals mean. Months of training were part of the story (S9). That training is not a failure; it is exactly what one would predict from the brain's flexibility. When the retina sends a novel code, the visual cortex does not instantly know how to interpret it. But the cortex is not a rigid decoder; it can learn new mappings, especially when the input becomes consistent and behaviourally relevant.

This is where retinal optogenetics becomes, unexpectedly, a window onto a general principle about brain interventions. In the mood circuits discussed earlier, the hope is often that the "right" stimulation will directly normalize a malfunctioning network. In the eye, the intervention is more plainly an interface: a new channel whose usefulness depends on the brain's ability to adapt. The boundaries between treatment and training blur. The patient is not a passive recipient of a fix; the patient becomes an active partner in teaching their brain to use a new sensory instrument.

Why, then, is the eye not already a solved problem? Because the retina's accessibility does not eliminate the hard parts; it simply makes them addressable.

Spatial resolution is one challenge. Normal vision depends on an extraordinarily dense sampling of the visual scene. If opsin expression is limited to certain cells or areas, and if stimulation is delivered through a device with limited resolution, the resulting percept may be coarse. That coarseness can still be meaningful—detecting the presence of an object, distinguishing a doorway from a wall, locating a moving shape—but it is far from reading ordinary print or recognizing faces effortlessly. There is also the question of what kind of retinal cell to target. Retinal ganglion cells are attractive because they are the output neurons, so making them light-sensitive can drive a signal directly to the brain. Yet ganglion cells are diverse; they carry different feature channels (motion, contrast, edges). Altering them risks changing the balance of information the brain receives. Targeting upstream cells like bipolar cells can, in principle, preserve more of the retina's internal processing, but it poses its own delivery and specificity challenges. These are design decisions, not mere technicalities.

Another challenge is variability. Viral delivery does not produce perfectly uniform expression across cells. Some cells express strongly; others weakly; some not at all. That heterogeneity changes how the retina responds to light patterns and complicates the design of the encoding device. It also complicates outcome prediction: two patients with the same diagnosis may not have the same surviving retinal architecture, the same receptive field organization, or the same capacity for cortical adaptation. This variability is part of why, as of mid-2025 reporting, peer-reviewed clinical optogenetics literature remained limited and clinical "firsts" should be framed as early feasibility rather than broad applicability (S15). A single proof-of-concept can demonstrate possibility without yet defining reliability.

Still, the broader significance of the eye-first story is not a triumphalist claim that optogenetics has arrived in the clinic. It is a demonstration of how a technology becomes medicine: by finding the place where physics, biology, and measurement line up well enough that careful human testing is ethically and technically imaginable. The retina is that place. Photons already belong there. Clinicians can see what they are doing. Engineers can build devices that interact with it. And patients can tell you, in the most direct language available, whether the world has become more navigable.

In that sense, restored sight through optogenetics is not just about vision. It is about earning a second chance at communication between the outside world and the nervous system—a conversation carried by light, translated into spikes, and gradually turned into meaning through practice.

11.4 The Price of Precision

"Precision" is a word that sounds like pure virtue. In medicine it suggests the opposite of blunt force: fewer side effects, clearer mechanisms, a better fit between problem and remedy. Optogenetics, with its promise of targeting particular neurons and controlling them with light, seems almost built to embody that ideal. Yet the closer a technology comes to being used in human beings, the more precision acquires a price tag. The bill is paid in invasiveness, uncertainty, time, regulation, and ethics. It is also paid in the most ordinary currency of all: the gap between what we intend to do to a circuit and what actually happens inside living tissue.

Start with a practical constraint that tends to get lost in the romance of light: light spreads. In brain tissue it scatters and fades, illuminating a volume rather than a point. A fibre tip placed in one location can bathe many nearby cells, including axons passing through from elsewhere. In the retina, illumination is easier to deliver but still not infinitely controllable; safety limits on intensity and wavelength are explicit design constraints that can force reliance on devices and shape achievable acuity and performance (S10). In either setting, the physical reality is that "a little bit of light" is not a scalpel. It is closer to a flashlight beam whose edges blur.

Biology adds its own blur. Opsin expression is rarely uniform. Viral delivery can lead to heterogeneous expression across cells: some take up the vector and express strongly, some weakly, some not at all. That heterogeneity is not just a nuisance; it changes the effective circuit manipulation. A stimulation protocol chosen on paper may overdrive a subset of cells and barely affect others. In a complex network, that unevenness can tilt activity in unanticipated ways, because circuits are not passive wiring diagrams but adaptive systems with inhibition, recurrent loops, and homeostatic responses. Precision is limited by physics and biology: light spread and scattering, heterogeneous opsin expression, and network indirect effects complicate the claim that optogenetics "targets exactly one thing" (S12).

Those network indirect effects deserve special emphasis, because they are an ethical issue as much as a technical one. A clinician might aim to increase activity in a pathway thought to support motivation, but the immediate result could be a cascade: inhibitory interneurons recruit, downstream regions rebound, other neuromodulators change, and the system finds a new equilibrium that is not the one intended. A person is not a preparation on a microscope stage; they carry memories, habits, and vulnerabilities. In the mood domain, where symptoms already involve self-evaluation and agency, unintended changes raise unusually sensitive questions. If an intervention alters irritability, risk-taking, or the sense of self, is that a side effect, a trade-off, or a different kind of harm?

Gene delivery makes these questions sharper. Translating optogenetics to humans generally requires introducing genes that encode an opsin, typically delivered by a viral vector such as AAV. Human translation adds gene-therapy burdens: immune responses to vector and/or opsin, dosing constraints, and long-term expression safety concerns (S11). Unlike a pill that can be stopped, gene delivery is not easily reversed. Even if expression wanes over time, it can persist for long periods, and that persistence changes the moral arithmetic. When a treatment is durable, patients are asked to consent not only to immediate risks but to a future in which an engineered protein may remain part of their tissue, with unknown interactions as the body ages.

Safety is not a single axis here; it has distinct dimensions that can be traded off against each other. Phototoxicity differs from excitotoxicity driven by opsin activation, and immune mitigation strategies (vector engineering, dosing, pharmacology) are central (S14, S11). Phototoxicity refers to damage caused by light itself, including heating or photochemical injury, a particular concern in delicate tissues like the retina. Excitotoxicity refers to damage caused by driving neurons too hard: excessive depolarization and firing can stress cells metabolically and disrupt ionic balances, potentially harming neurons or their synapses. One can imagine reducing light intensity to protect tissue from phototoxicity, but then compensating by using a more light-sensitive opsin or increasing expression, which might increase the risk of driving cells into unhealthy regimes. Or one can imagine reducing excitotoxicity by spreading stimulation out, which may reduce precision and require higher total light power to achieve an effect. Each adjustment solves one problem by leaning into another.

A concrete example of this "precision paradox" arises naturally in the eye. The retina is accessible and can be illuminated directly, but safe illumination limits and wavelength choices constrain what can be delivered (S10). That is one reason current approaches have relied on external devices such as goggles: the device can intensify and shape the stimulus while staying within conservative safety margins, and it can choose wavelengths appropriate for the opsin and the tissue. Yet the very existence of the device highlights the paradox. A biological intervention designed to be precise still needs an engineered workaround to be usable. Precision at the cellular level does not automatically translate into precision at the perceptual level.

The deeper brain raises a more stark version of the same paradox. To stimulate specific deep structures with light, hardware generally must be implanted. Implants bring surgical risk, infection risk, and long-term maintenance issues. They also introduce practical constraints: hardware placement is never exact to the micron, and small placement differences can change which fibres and cell bodies are stimulated. More precision may demand more invasive delivery; less invasiveness may force broader, less controllable effects. A technology can be exquisitely specific in principle and still difficult to deploy safely in practice.

These are the biomedical costs. Ethics adds a second ledger.

The first ethical question is consent, and it is not as straightforward as signing a form. Consent in invasive neurotechnology involves understanding uncertainty, and uncertainty is not evenly distributed. A person with advanced retinal degeneration may be willing to take risks for a chance at partial visual recovery. A person with severe depression who has not responded to existing treatments may be similarly willing. But depression itself can distort hope, risk perception, and self-worth. When suffering is the very condition that makes a person eligible for an intervention, the line between autonomous choice and desperate grasping becomes delicate.

This is not a reason to deny experimental treatments to people in need; it is a reason to take special care with how hope is offered. Optogenetics, perhaps more than many interventions, invites vivid mental images: "turning on" a circuit, "restoring" a function, "reprogramming" a network. Such language can inflate expectations in ways that are not merely disappointing but ethically corrosive. It can subtly shift the burden onto the patient: if the brain is a controllable machine, then failure begins to look like personal malfunction rather than the normal outcome of uncertain biology.

The second ethical question is identity. There is a long philosophical tradition of asking whether we are our brains, but in the clinic the question becomes immediate and practical. If a mood intervention changes what a person wants, enjoys, fears, or values, does that count as treatment success? Sometimes the answer is clearly yes: a person no longer trapped in suicidal despair is not "less themselves" in any meaningful moral sense. Yet there are gray zones. Treatments that alter motivation can also alter impulsivity. Treatments that reduce negative affect can, in some individuals, flatten emotional range. A person might gain energy but lose caution; gain sociability but lose patience. These are not hypothetical worries; they are the kinds of trade-offs that have appeared in various forms in neuromodulation history, and optogenetics would inherit the same ethical terrain, intensified by the permanence of gene delivery.

The third ethical question is access. A sophisticated therapy that combines viral manufacturing, surgical or ophthalmic delivery, specialized devices, and long-term rehabilitation is unlikely to be cheap. If such interventions eventually become effective, who gets them? The answer is not simply a matter of insurance policy; it shapes how society values disability and mental illness. A world in which only wealthy patients can obtain a "precision" therapy is a world that turns precision into a new axis of inequality.

Policy and regulation are often treated as obstacles to innovation, but in neurotechnology they are also a kind of social memory. They encode lessons learned the hard way about harm, hype, and the slow tempo of evidence. In optogenetics, oversight, consent, and dual-use concerns influence how the technology is perceived and governed (S18). Dual-use here means that tools designed for therapy and research can also be imagined for coercion or enhancement. A technique capable of altering behaviour by targeting circuits will inevitably attract speculative fears about mind control. Those fears can be exaggerated, but dismissing them entirely is unwise. The ethical response is not paranoia; it is clarity about what is feasible, what remains science fiction, and what safeguards are built into research practice.

A more mundane, but equally important, regulatory challenge is evidence. In the laboratory, a striking behavioural effect in a rodent can be obtained in a controlled environment, with carefully chosen stimulation patterns, in animals of similar age and genetics. In human populations, variability is the rule. Regulators and clinicians must decide what counts as meaningful benefit. In vision restoration, detecting a flash is not the same as navigating a crowded street. In depression, a change on a rating scale is not the same as being able to return to work or reconnect with family. When an intervention is invasive and potentially durable, the threshold for benefit must be high enough to justify risk, and the evidence must be strong enough to rule out placebo effects, regression to the mean, and the powerful influence of expectation.

A historical anecdote makes this logic emotionally vivid. In the early days of psychosurgery, procedures such as lobotomy were promoted as medical miracles, sometimes on the basis of dramatic short-term changes in agitation or distress. The long-term costs, in personality and cognition, were often minimized or reframed as acceptable. That era left a scar on psychiatry and on public trust. Modern neuromodulation is ethically and scientifically far removed from that history, with far stricter oversight and a very different understanding of brain function. Still, the lesson persists: interventions that can change behaviour can be misused when desperation meets charisma, or when institutions reward rapid "breakthroughs" more than careful follow-up.

Optogenetics, at its best, encourages the opposite temperament. It trains researchers to specify exactly what they did: which cell type, which projection, which temporal pattern, and in what state of the animal. It also forces honest confrontation with the distance between manipulation and outcome. If a stimulation produces a behavioural change, the change may be mediated by indirect network effects; if it fails, the failure may reflect expression variability, insufficient light, or state dependence rather than a wrong hypothesis. That discipline can make the technology ethically safer by discouraging overconfident claims.

Yet discipline does not remove the underlying moral burden. When optogenetics becomes therapy, it is no longer only a way of learning about brains. It becomes a way of changing lives, and therefore a way of distributing risks and opportunities. The price of precision is not merely the cost of a device or a manufacturing pipeline. It is the obligation to match a powerful tool with an equally powerful culture of restraint: careful evidence, candid communication, and a respect for the fact that the brain is not a target to be hit, but a person to be helped.

11.5 Reading While Writing

A clean optogenetic experiment has the feel of a decisive question. Shine light into a defined pathway and see what behaviour changes. But that decisiveness can be theatrical. Inside the brain, the light-driven command is not a sentence delivered to a silent audience; it lands in the middle of a crowded conversation. Some neurons respond strongly, some weakly. Some are already close to firing, others are held down by inhibition. The same pulse train can be swallowed on one day and amplified on another. That is why the most mature optogenetic work has started to treat stimulation as only half of the act. The other half is reading out what the circuit actually did in response, moment by moment.

The phrase "reading while writing" captures a change in scientific attitude. Writing means intervening: stimulating or silencing neurons. Reading means measuring: recording spikes, imaging activity, tracking behaviour, watching physiology. In everyday life, writing without reading is how you send a message into the void; you have no idea whether it arrived or what it caused. In the laboratory, writing without reading invites a quieter failure: you believe you controlled a circuit because you delivered light, but you never confirm what changed in the living network.

The need for readout is not merely methodological hygiene; it reshapes what counts as an explanation. A circuit story that sounds crisp in words often dissolves when measurement is added. "Activate this node to produce resilience" is a slogan. A more defensible claim specifies what was activated (cell type and projection), what pattern was delivered (timing), what state the brain was in, and what activity changes followed across the network. Optogenetics is strongest when paired with readouts that validate what stimulation actually did, because expression and network context can diverge from intent (S12, S13). Without those readouts, an optogenetic result can be real and yet poorly understood: a behavioural change with an unknown internal route.

A historical anecdote helps put this in perspective. When physiologists first began to stimulate nerves electrically in the nineteenth century, they quickly learned that the same stimulus could produce different outcomes depending on the condition of the tissue: temperature, fatigue, prior activity, the ionic environment. That lesson did not slow the field; it made it scientific. It pushed experimenters toward quantitative measurement and toward the recognition that a biological system is not simply a wire but a device with state. Modern optogenetics is replaying that maturation at a higher level of complexity. The intervention is more specific, but the need to measure is more urgent, because the network has more ways to respond indirectly.

The dream version of reading while writing is "all-optical" control. Instead of delivering light through one channel and recording through another, one uses light both to perturb neurons and to observe the resulting activity across many cells. The basic trick is to pair two kinds of light-sensitive molecules. One is an opsin, used to drive or inhibit neurons. The other is an optical activity reporter, often a calcium indicator: a protein whose fluorescence changes when intracellular calcium rises, which tends to happen when neurons fire action potentials. Calcium imaging does not record spikes with perfect fidelity, but it provides an extraordinary window into population dynamics: which cells are active, who is synchronized with whom, how patterns sweep across tissue.

In a typical all-optical experiment, a microscope images hundreds or thousands of neurons while patterned light targets selected cells or regions. Patterned light can be shaped in space and time, allowing the experimenter to stimulate one ensemble while observing activity in many others. Two-photon microscopy, which uses infrared light and nonlinear optics, can restrict excitation to a tiny focal volume, making it possible to image and stimulate deeper in tissue with reduced out-of-focus effects. The details are technical, but the conceptual leap is simple: instead of guessing what the stimulation did, you watch the circuit answer back.

That dialogue is especially valuable in mood-related networks, where timing dependence and plasticity make simplistic cause-and-effect stories brittle. Interventions can have acute effects and longer-lasting synaptic consequences, strengthening the case for protocols that respect state and history rather than treating the brain as a fixed target

(S2). In the laboratory, that means noticing whether a stimulation pushes the network into a new regime that persists after light is off, or whether it produces only a transient ripple. It means asking whether a behavioural change corresponds to a measurable shift in coordinated activity across prefrontal, limbic, and reward circuits, rather than a vague "increase" or "decrease" in one region.

Reading while writing becomes even more powerful when it turns into feedback control. In the simplest optogenetic designs, light is delivered on a schedule chosen in advance. That is "open-loop" control: the output of the system does not influence the input. Closed-loop optogenetics integrates measurement and stimulation so that the intervention depends on the current neural state, improving causal tests and pointing toward more adaptive therapeutic strategies (S13). A closed-loop system might detect a particular pattern—a burst, an oscillatory phase, a signature of a state transition—and stimulate only when that pattern appears. Or it might adjust stimulation intensity in real time to achieve a target activity level. The intervention becomes conditional, like steering rather than pushing.

This is not merely an engineering flourish; it changes the scientific question from "can I force the circuit to do something?" to "can I guide the circuit when it starts to do the wrong thing?" That second question is closer to many clinical realities. Depression, anxiety, addiction, and compulsive disorders often involve episodes: spirals of rumination, waves of craving, sudden avoidance, panic that snowballs. An intervention delivered continuously may be unnecessary for long stretches and insufficient at the critical moment. A state-aware intervention—one that listens for the beginning of a spiral—could, in principle, do less overall while doing more where it counts.

Closed-loop designs also put pressure on a comforting illusion: that optogenetics is inherently clean. If identical light pulses produce different outcomes across animals, or across days in the same animal, the variability can be deeply informative. It may reflect changes in brain state: stress hormones, sleep debt, arousal, ongoing oscillations. It may reflect biological variability in opsin expression or in the network's compensatory responses. It may reflect differences in what the animal has learned. A closed-loop approach does not remove that complexity; it tries to work with it, using measurement to decide when and how to intervene rather than insisting on a one-size schedule.

The logic echoes a lesson from the retina, where the "readout" is not only a measurement device but the patient's own learning. In retinal optogenetic therapy, functional gains rely on training to interpret a new sensory code produced by the device and the transduced retina (S9). That is reading while writing in a different sense. The goggles deliver a patterned stimulus to the retina—writing to the nervous system—and the brain gradually learns to read that stimulus as meaningful information. The feedback loop is behavioural rather than electronic. The person moves, reaches, searches, and the brain updates its interpretation of the new signals. Rehabilitation is not an optional add-on; it is how the intervention becomes useful.

That parallel is easy to miss if one treats "closed-loop" as a purely technical term. In reality, the nervous system is always a closed loop with the world. Actions change sensory input. Expectations change attention. Learning changes what counts as salient. Any brain intervention that alters perception, motivation, or mood will interact with these loops. The lesson from retinal therapy is simply more visible because the loop is concrete: a device projects patterns into the eye, the person practises, and perception changes. Mood interventions are looped too, but in a more elusive way: a shift in motivation changes social engagement, which changes reward exposure, which changes mood, which changes motivation. Reading while writing, at its most ambitious, would incorporate those loops rather than pretending they are noise.

All-optical and closed-loop methods also help with a philosophical problem that shadows brain science: when does a manipulation become an explanation? A single optogenetic effect can tempt one into thinking a circuit has been "found." But a true explanation should let you predict and control outcomes across contexts, and should tell you why a given manipulation works when it does and fails when it fails. Simultaneous readout helps convert a manipulation into a model. If stimulating a pathway produces a behavioural shift only when a certain network signature is present, that signature becomes part of the explanation. If stimulation reorganizes activity across a distributed network in a reproducible way, that reorganization can be tested, refined, and linked to mechanism.

In practical terms, reading while writing has encouraged a new kind of experiment: not just "activate region A" but "activate a defined ensemble in region A and measure how regions B and C respond, then activate a different ensemble and compare the network trajectory." The emphasis shifts from single-node control to network dynamics. For mood, that seems especially appropriate. Sadness, motivation, vigilance, and hope do not reside in one spot. They are emergent states of a whole organism negotiating its world. A technology that can both perturb and observe distributed patterns is, at minimum, better matched to the phenomenon.

There is also a moral payoff to this approach. Earlier, the ethical burden of precision came partly from uncertainty: not knowing exactly what a stimulation did inside a brain, and therefore not knowing what side effects might follow. Better readout does not eliminate risk, but it makes the risks more legible. It allows researchers to say not only "we delivered light" but "we observed these changes in circuit activity, and we can stop stimulation when the circuit leaves the target regime." In a field where hopes can outrun data, the ability to show what happened in the tissue is a form of honesty.

The long-term future of optogenetics in medicine remains uncertain, especially for psychiatric disorders where delivery and measurement are difficult. Yet the intellectual direction is clear. The most revealing optogenetic experiments are no longer those that merely flip a switch, but those that treat the brain as a responsive system: one that must be listened to, not just commanded. When intervention becomes a conversation—measure, perturb, measure again, adjust—the questions change from "where is the problem?" to "what pattern is the system trapped in, and what nudge helps it regain flexibility?" That is a different kind of precision, aimed not at a single molecule or a single location, but at restoring the brain's ability to move gracefully among its own states.

www.ingramcontent.com/pod-product-compliance
Lightning Source LLC
LaVergne TN
LVHW010712061126
851857LV00014B/1144
9798240804595